# 2010 IEEE Workshop on Microelectronics and Electron Devices

# (WMED 2010)

Boise, Idaho, USA
16 April 2010

IEEE Catalog Number: CFP10564-PRT
ISBN: 978-1-4244-6572-9

**Copyright © 2010 by the Institute of Electrical and Electronic Engineers, Inc**
**All Rights Reserved**

*Copyright and Reprint Permissions*: Abstracting is permitted with credit to the source. Libraries are permitted to photocopy beyond the limit of U.S. copyright law for private use of patrons those articles in this volume that carry a code at the bottom of the first page, provided the per-copy fee indicated in the code is paid through Copyright Clearance Center, 222 Rosewood Drive, Danvers, MA 01923.

For other copying, reprint or republication permission, write to IEEE Copyrights Manager, IEEE Service Center, 445 Hoes Lane, Piscataway, NJ 08854. All rights reserved.

***This publication is a representation of what appears in the IEEE Digital Libraries. Some format issues inherent in the e-media version may also appear in this print version.*

IEEE Catalog Number:      CFP10564-PRT
ISBN 13:      978-1-4244-6572-9
ISSN:      1947-3842

**Additional Copies of This Publication Are Available From:**

Curran Associates, Inc
57 Morehouse Lane
Red Hook, NY 12571 USA
Phone:     (845) 758-0400
Fax:     (845) 758-2633
E-mail:     curran@proceedings.com
Web:     www.proceedings.com

# 2010 IEEE Workshop on Microelectronics and Electron Devices (WMED 2010)

Boise, Idaho, USA
16 April 2010

| | |
|---|---|
| IEEE Catalog Number: | CFP10564-POD |
| ISBN: | 978-1-42446-572-9 |

# 2010 IEEE WMED – Table of Contents

Welcome to the 2010 IEEE WMED

2010 IEEE WMED Management Committee

WMED-2010 Technical Program

WMED 2010 Contributed Talks:

## Keynote and Invited Speakers Session

CeNSE: The Central Nervous System for the Earth ............................................................................... 3
Peter G. Hartwell, Hewlett-Packard Laboratories
Challenges and Innovations in Nano-CMOS Transistor Scaling ....................................................... 4
Tahir Ghani, Intel Corporation
Challenges and Opportunities Moving from 2D Chips to 3D Chips .................................................. 5
James Jian-Qiang Lu, Rensselaer Polytechnic Institute
Hydrogenated Silicon (Si:H) Thin Film Solar Cells .......................................................................... 6
C.R. Wronski, Pennsylvania State University
Carbon based Nanomaterials as Interconnects and Passives for Next-Generation VLSI and 3-D ICs ....................... 7
Kaustav Banerjee, University of California, Santa Barbara
Practical Semiconductor Reliability ...................................................................................................... 8
Todd Marquart, Micron Technology

## Technical Presentations

## Process and Devices Session

Modified Floating Gate and IPD Profile for Better Cell Performance of
Sub-50nm NAND Flash Memory ....................................................................................................... 11
Jennifer Lequn Liu, Fernando Gonzalez, Y. Jeff Hu, Micron Technology, Inc.; Jixin Yu, Charan Srinivasan,
and Ervin Hill, Intel Corporation

Study of Carrier Mobility of Low-Energy, High-Dose Ion Implantations using Continuous
Anodic Oxidation Technique/Differential Hall Effect (CAOT/DHE) Measurements ....................... 15
Shu Quin, Y. Jeff Hu, Allen McTeer, Micron Technology, Inc.; Si Prussin, and Jason Reyes, University
of Los Angeles

Discrete Test Structure Device Degradation Analysis and Correlation to NAND
Flash Circuit Operation ...................................................................................................................... 19
Jasper Gibbons, Puneet Sharma, Steve Porter, Jim Fulford, Praveen Vaidyanathan, Sheryll De Guzman,
Pratap Murali, and Ken Marr, Micron Technology, Inc.

A Comprehensive Study on Nanomechanical Properties of Various $SiO_2$-based Dielectric Films ......................... 22
Guohua Wei, Song Varghese, Kevin Beaman, Irina Vasilyeva, Tom Mendiola, Andrew Carswell,
David Fillmore, and Shifeng Lu, Micron Technology, Inc.

A Novel Depletion Mode High Voltage Isoloation Device ................................................................ 26
Vladimir Mikhalev, and Michael Smith, Micron Technology, Inc.

Fullband Study of Ultra-scaled Electron and Hole SiGe Nanowire FETs ........................................ 29
Abhijeet Paul, Saumitra Mehrotra, Methieu Luisier, and Gerhard Klimeck, Purdue University

## Circuits Session

Continuous-Time/Discrete-Time (CT/DT) Cascaded Sigma-Delta Modulator for
High Resolution and Wideband Applications ..................................................................................... 33
Ali Mesgarani, University of Idaho; Khosrow H. Sadeghi, Sharif University of Technology;
and Suat U. Ay, University of Idaho

All Digital Multiplying DLL Using Precision Digital Delay Line as DCO ........................................ 37
Seong-Hoon Lee, Micron Technology, Inc.

Main Memory with Proximity Communication, A Wide I/O DRAM Architecture ................................................ 40
Qawi Harvard, R. Jacob Baker, Boise State University; and Robert Drost, Sun Microsystems Laboratory

A Low Noise Low Power DC Coupled Sensor Amplifier with Offset Cancellation ................................................ 44
Hari Krishnan Krishnamurthy, Dirk Robinson, Dave M. Rector, and Geroge S. La Rue,
Washington State University

Integration of a New Column-Parallel ADC Technology on CMOS Image Sensor ................................................ 48
Fan Z. Nelson, and Suet U. Ay, University of Idaho

Gain Error Correction for CMOS Image Sensor using Delta-Sigma Modulation ................................................ 52
Kuangming Yap, and R. Jacob Baker, Boise State University

Poster Presentations and Abstracts ................................................ 57

Advanced Call for Papers – WMED 2011 ................................................ 59

# A Letter from the WMED Chairman

Welcome to the eighth IEEE WMED. The WMED (Workshop on Micro-Electronic Devices) is an IEEE professional society event hosted by the Electron Device Society Section 6. We are excited to be back again on the Boise State University campus at the Student Union Building for this years workshop.

The previous year, 2009, was an extremely difficult year for our industry. Locally the Treasure Valley was not spared the affects of the industry downturn and in the last year has experienced layoffs, plant closings and the contraction or disappearance of many local hi-tech companies. Therefore I'm happy to still see a strong interest in the workshop again this year, proving that the workshop is still filling a regional educational need in our profession and community. The WMED committee and I believe we have assembled a program today that will be both relevant and educational for all.

I must thank our invited speakers and contributors for sharing their knowledge and expertise with us today. In addition I must also thank our generous sponsors for their financial support, in particular Micron Foundation, the Boise State University ECE department, and the local IEEE and IEEE-EDS chapters. The workshop is indebted to the hard work and dedication of our volunteer staff who lend their time and talent to this endeavor each year. I want to thank the WMED University Advisory Board for their helpful suggestions and support this year. I must also thank you, our participants here today, both the students and professionals, because without your interest and support each year there would be no workshop. For me personally it is an honor to be your host today.

Today we have invited speakers from around the country who are outstanding in their fields. For the second year in row we are proud to have another IEEE Fellow, Tahir Ghani of Intel, speak at our workshop. He will address the challenges and implications of device scaling in the nanometer regime. This year we are also pleased to have two speakers who are recognized in the industry for their innovation. Peter Hartwell of HP Labs has invented a new class of MEMS accelerometer that is more sensitive as those found in consumer devices. This new sensor forms the key part of HP's CeNSE initiative and is the focus of his talk today. C.R. Wronski from Penn State is the co-inventor of one type of solar cell and has researched optoelectronic devices since the early 1960s. His talk today centers on the evolution of the technology of silicon thin film solar cells. James Lu of RPI is a researcher in the topic of 3D semiconductor scaling. The 3D approach for semiconductors is expected to extend the scaling of integrated circuits in the future and is currently of great interest and activity in the industry.

A valuable part of the workshop is our technical tutorials. We have choice of two tutorials today. Our tutorial on semiconductor reliability from Todd Marquart of Micron reviews the methods and techniques needed to analyze the reliability of modern nano-scale semiconductor processes and the new challenges that are facing reliability engineers as we continue to shrink the devices. Our other tutorial from Kaustav Banerjee of UC Santa Barbara will focus on the future of interconnect scaling and the possibility of moving to new nano-scale materials that offer better properties than the copper wiring that is currently used in today's integrated circuits. Following the invited speakers and tutorial sessions will be two parallel technical sessions and a poster session. These sessions feature a combination of academic and industry research.

During the workshop IEEE WMED provides a unique program for local area high school students. The WMED high school program introduces students to the field of engineering as a career choice and features talks and presentations highlighting the engineering experience from several different perspectives. These include a technical talk from Dean Klein the VP of Memory System Development for Micron and a forum for discussing the transition from high school to college. Several local IEEE members are supporting this event along with Micron's K-12 Manager, Alecia Baker.

William Kueber
IEEE WMED 2010 General Chair

# IEEE WMED 2010

This workshop owes a debt of gratitude to all the people and sponsors that have donated their resources, time and talent to make this workshop possible this year.

## IEEE WMED 2010 Administrative Committee

William Kueber – General Chair
Vishwanath Bhat - Technical Chair
Prashant Raghu - Publications Chair
Huy Le - Finance Chair
Kyle Campbell - Publicity Chair
Tim Hollis - Program Chair
David Wells - Registration Chair
Sachin Joshi - Webmaster
Mark Eyolfson - High School Program Chair
Alecia Baker - High School Program Advisor
Shyam Surthi, Steve Groothuis, Fernando Gonzalez  - Senior Advisors
Jaydip Guha, Ionicia Dembi, Leslie Bird, Rohit Rai, Chris Anderson – Volunteers
Hari Naidu - Photographer

## IEEE WMED 2010 University Advisory Board

Prof. Herb Hess, University of Idaho
Prof. Jim Browning, Boise State University
Prof. Jacob Baker, Boise State University

## IEEE Electron Devices Society (EDS) Boise Chapter Officers

Shyam Surthi - Chapter Chair and Secretary
Jaydip Guha - Vice Chair
Huy Le - Treasurer
Tony Liu - Membership Chair
Kyle Campbell - Publicity
Jim Browning - University Liaison
Steve Groothuis, Gurtej Sandhu – Advisors

## IEEE WMED2010 Sponsors

Micron Foundation
Boise State University College of Engineering
IEEE Boise Section
IEEE-EDS Region 6

## Manuscript Reviewers for WMED2010

Adam Johnson, Alan Mondada, Anish Khandekar, Bhaskar Srinivasan, Deepak Raghu, Dragos Dimitriu, Fernando Gonzalez, Glen Hush, Howard Kirsch, Jake Baker, Jaydeb Goswami, Jaydip Guha, Jen Sigman, Kaveri Jain, Nikolay Mirin, Pavan Aella, Puneet Sharma, Randy Koval, Sachin Joshi, Sampat Ratnam, Satya Saripalli, Shashi Hegde, Srivardhan Gowda, Suraj Mathew, Suresh Ramakrishnan, Tim Hollis, Vassil Antonov, Vikram Bollu, Vinay Nair, Vishal Sipani, Wayne Huang

## WMED-2011

If you would be interested in volunteering to help for WMED or another IEEE activity in Boise please contact the general chair or the EDS chapter chair (billrk@ieee.org, or ssurthi@micron.com )

# IEEE WMED 2010 Technical Program

## Friday, April 16, 2010 8:00AM-7:00PM

| | |
|---|---|
| 8:00AM | **Check In and Door Registration**<br>*Continental Breakfast in Jordan-D* |
| 8:30AM | **Welcome to WMED2010**<br>*Jordan-D Ballroom* |
| 8:45AM | **Invited Talk: "CeNSE: The Central Nervous System for the Earth"**<br>*Dr. Peter Hartwell, Hewlett-Packard Labs, Palo Alto*<br>*Jordan-D Ballroom* |
| 9:45AM | **Break & Poster Setup** |
| 10:00AM | **Invited Talk: " Challenges and Innovations in Nano-CMOS Transistor Scaling"**<br>*Dr. Tahir Ghani, IEEE Fellow, Intel Corporation*<br>*Jordan-D Ballroom* |
| 11:00AM | **Invited Talk: "Challenges and Opportunities Moving from 2D to 3D Chips"**<br>*Dr. James Lu, RPI*<br>*Jordan-D Ballroom* |
| NOON | **Buffet Luncheon**<br>*Provided by the Workshop*<br>*Jordan-AB* |
| 1:00PM | **Invited Talk: "Hydrogenated Silicon (Si:H) Thin Film Solar Cells"**<br>*Dr. C.R. Wronski, Penn State University.*<br>*Jordan-D Ballroom* |
| 2:00PM | **Invited Tutorials**<br>**Tutorial A: "Carbon based Nanomaterials as Interconnects and Passives for Next-Generation VLSI and 3-D ICs", Jordan-D**<br>*Dr. Kaustav Banerjee, UC Santa Barbara*<br>**Tutorial B: "Practical Semiconductor Reliability": Jordan-C**<br>*Dr. Todd Marquart, Micron Technology*<br>*(Tutorials are two hours and parellel)* |
| 4:00PM | **Break** |
| 4:15PM | **Paper Sessions**<br>**Process and Devices: Jordan-D**<br>**Solid State Circuits: Jordan-C**<br>*(Sessions are parellel)* |
| 5:45PM | **Poster Presentation and Reception**<br>*Jordan-D Ballroom* |

*This workshop is receiving technical co-sponsorship support from the IEEE Electron Device Society*

# Contributed Talks: Parallel Sessions (4:15PM-5:45PM)

## Process and Devices Session

**4:15p** . . . . . . . . **Modified Floating Gate and IPD Profile for Better Cell Performance of Sub-50nm NAND Flash Memory**
Jennifer Lequn Liu, Fernando, and Y. Jeff Hu, Micron Technology, Inc., Jixin Yu, Charan Srinivasan, Ervin Hill, Intel Corporation

**4:30p** . . . . . . . . **Study of Carrier Mobility of Low-Energy, High-Dose Ion Implantations using Continuous Anodic Oxidation Technique/Differential Hall Effect (CAOT/DHE) Measurements**
Shu Quin, Y. Jeff Hu, Allen Mcteer, Si Prussin, Jason Reyes, Micron Technology, Inc.

**4:45p** . . . . . . . . **Discrete Test Structure Device Degradation Analysis and Correlation to NAND Flash Circuit Operation.**
Jasper Gibbons, Puneet Sharma, Steve Porter, Jim Fulford, Praveen Vaidyanathan, Sheryll De Guzman, Pratap Murali, Micron Technology, Inc.

**5:00p** . . . . . . . . **A Comprehensive Study on Nanomechanical Properties of Various $SiO_2$-based Dielectric Films.**
Guohua Wei, Song Varghese, Kevin Beaman, Irina Vasilyeva, Tom Mendiola, Andrew Carswell, David Fillmore, and Shifeng Lu, Micron Technology, Inc.

**5 :15p** . . . . . . . . **A Novel Depletion Mode High Voltage Isoloation Device**
Vladimir Mikhalev, Michael Smith, Micron Technology, Inc.

**5:30p** . . . . . . . . **Atomistic Study of Ultra-scaled Electron and Hole SiGe Nanowire FETs**
Abhijeet Paul, Saumitra Mehrotra, Methieu Luisier, Gerhard Klimeck, Purdue University

## Circuits Session

**4:15p** . . . . . . . . **Continuous-Time/Discrete-Time (CT/DT) Cascaded Sigma-Delta Modulator for High Resolution and Wideband Applications**
Ali Mesgarani, Suat U. Ay, University of Idaho, Khosrow H. Sadeghi, Sharif University of Technology.

**4:30p** . . . . . . . . **All digital Multiplying DLL Using Precision Digital Delay Line as DCO**
Seong-Hoon Lee, Micron Technology, Inc.

**4:45p** . . . . . . . . **Main Memory with Proximity Communication, A Wide I/O DRAM Architecture**
Qawi Harvard, R. Jacob Baker, Boise State University, Robert Drost, Sun Microsystems Laboratory

**5:00p** . . . . . . . . **A Low Noise Low Power DC Coupled Sensor Amplifier with Offset Cancellation**
Hari Krishnan Krishnamurthy, Dirk Robinson, Dave M. Rector, Geroge S. La Rue, Washington State University.

**5:15p** . . . . . . . . **Integration of a New Column-Parallel ADC Technology on CMOS Image Sensor**
Fan Z. Nelson, Suet U. Ay. University of Idaho

**5:30p** . . . . . . . . .**Gain Error Correction for CMOS Image Sensor using Delta-Sigma Modulation**
Kuangming Yap, R. Jacob Baker, Boise State University

## Poster Session

**5:45p** . . . . . . . . **Enhanced Optical Transmission in Hexagonal Plasmonic Crystals**
A. English, L. Lowe, and W. Kuang, Boise State University

**A Compact Delay-Locked Loop for Multi-Phase Non-Overlapping Clock Generation**
Chris Gagliano and R. Jacob Baker, Boise State University

**A 16-bit 500KSps Low Power SAR ADC**
Kun Yang and George S. La Rue, Washington State University

**Design of an On-Chip Quasi-Resonant Fixed Frequency Buck DC-DC Power Converter**
Lucas A Wells, University of Idaho

**Method of determination pattern placement errors (PPE) due to scanner lens aberration by using product circuit pattern with double patterning technology**
Maiko Uemura and Masato Shinohara, Micron Japan, Nishiwaki City, Hyogo, Japan

## Poster Session (con't)

5:45p . . . . . . . **Using Gate Voltage Sensitivity to Analyze Bvdss in NAND Periphery RESURF Devices**
Michael A. Smith, Micron Technology Inc

**Damage Engineering of Boron-Based Low Energy Ion Implantations on USJ Fabrications**
Shu Qin, Y. Jeff Hu, and Allen McTeer, Micron Technology Inc.

# KEYNOTE AND INVITED SPEAKERS SESSION

978-1-4244-6572-9/10 $26.00 © 2010 IEEE

# Invited Talk:
## "CeNSE: The Central Nervous System for the Earth"
### Dr. Peter G. Hartwell
### Hewlett-Packard Laboratories, Palo Alto, California

We stand on the threshold of the next phase of technology evolution, the era of the sensor. As the information age has progressed from computer to networks to the cloud our ability to store and share knowledge now impacts every aspect of our lives. The next step is to bring awareness to the information system through a distributed network of sensors.

HP's Central Nervous System for the Earth (CeNSE) is a meta-project involving multiple HP Labs focused on enabling a planetary system of a trillion nanoscale sensors and actuators embedded in the environment and the networks to exchange their information among analysis engines, storage systems and end users. CeNSE will revolutionize human interaction with the Earth in a manner as profound as the Internet revolutionized our interactions with other humans. The massive amounts of data harvested and subsequent actions to be taken will create an enormous increase in demand for computing systems and services. HP Lab's research covers all aspects required to create such a system from the nano-scale sensor node to wireless and photonic networks to storage and computation to data visualization, information theory and analysis. CeNSE will provide real-time monitoring of surroundings for mission critical business, environmental, health and safety applications, such as factory operations, merchandise tracking, large structure integrity, virus tracking, food safety, energy use, and many more.

### *Curriculum vitae*

Dr. Peter G. Hartwell is currently a senior researcher at Hewlett-Packard Laboratories in Palo Alto, California. As a member of the Information and Quantum Systems Lab, he is the lead of the Central Nervous System for the Earth (CeNSE) team developing a broad sensing system to bring environmental awareness to information technology infrastructure. CeNSE was selected one of 20 "World Changing Ideas" in the December 2009 issue of Scientific American. Peter has extensive experience in commercializing silicon MEMS products, working on advanced sensors and actuators, and specializes in MEMS testing techniques. He graduated from the University of Michigan in 1992 with a B.S.E in Materials Science and from Cornell University in 1999 with a Ph. D. in Electrical Engineering. He did brief post doctoral work at HP Labs before joining the staff in 2000. His work at HP has been documented in numerous technical papers, patents, and articles in publications such as The New York Times, Forbes, IEEE Spectrum, and EETimes.

# Invited Talk:
# "Challenges and Innovations in Nano-CMOS Transistor Scaling"
Tahir Ghani
Intel Fellow, Logic Technology Development
Intel Corporation


Starting at 90nm CMOS node, the industry started to experience significant barriers in achieving historical transistor performance gains through traditional dimensional scaling. Fortunately, the industry has responded positively to this challenge by implementing many innovations in device structure and materials to overcome traditional scaling barriers. Intel has been at the forefront in addressing these challenges by successfully driving transistor innovations from research phase to mainstream CMOS manufacturing. This talk will start by briefly outlining logic technology trends and discuss how innovations such as uniaxially-strained silicon and "HiK+Metal Gate" technologies, pioneered by Intel, have enabled dramatic performance / power enhancement for the recent CMOS nodes. I will next present how future innovations in devices and materials have the potential to address upcoming scaling concerns and highlight some fundamental challenges in implementing new innovations into mainstream CMOS technology


## *Curriculum vitae*

Tahir Ghani is an Intel Fellow and Director, Transistor Technology and Integration at Intel Corporation. Since joining Intel in 1994, he has led the teams responsible for developing some of the most significant changes in semiconductor industry and implementing them into mainstream CMOS manufacturing. Tahir co-led the team responsible for developing industry-first HiK+Metal Gate CMOS technology for Intel's 45nm technology node. and strained Silicon CMOS technology for Intel's 90nm technology node. As a member of Pathfinding team, he is currently leading transistor technology research and development for 15nm CMOS logic node. Tahir received his PhD in Electrical Engineering from Stanford University in 1994. He is a Fellow of IEEE.

# Invited Talk:
# "Challenges and Opportunities Moving from 2D Chips to 3D Chips"

James Jian-Qiang Lu

Department of Electrical, Computer, and Systems Engineering

Rensselaer Polytechnic Institute, luj@rpi.edu


Three-dimensional (3D) integration is an emerging technology, which vertically stacks and interconnects multiple materials, technologies, and functional components to form highly integrated micro-nano systems. This is expected to lead to an industry paradigm shift due to its tremendous benefits compared to 2D integration in performance, data bandwidth, functionality, heterogeneous integration and power. The key enabling technology is the formation of a massive number of small-sized Through-Strata-Vias (TSVs), which is the hottest research topic at present. However, there are a number of challenges associated with 3D integration, including evolution and integration of the complex fabrication technologies, integration architecture and design tools, thermal and mechanical constraints, yield and cost, and manufacturing infrastructure. After a brief introduction to the 3D integration, this paper will focus on some key challenges and potential solutions, research directions, as well as opportunities with 3D integration technologies.


*Curriculum vitae*

James Jian-Qiang Lu received his Dr.rer.nat. (Ph.D.) degree from Technical University of Munich in December 1995, and is currently an Associate Professor in Electrical Engineering at Rensselaer Polytechnic Institute (RPI), Troy, NY. At RPI, Dr. Lu has worked on 3D hyper-integration technology, design and applications since 1999, with focus on hyper-integration and micro-nano-bio interfaces for future chips. Dr. Lu has more than 200 publications in the areas from micro-nano-electronics theory and design to materials, processing, devices, integration and packaging (e.g., GaAs, GaN and Si devices, novel FETs, terahertz electronics, carbon-nanotubes, Si IC interconnects and integration for memory and processor). He and gave a number of invited presentations, seminars and short courses. He served as technical chair, workshop chair, session chair, panelist and panel moderator for many conferences. He is a senior member of IEEE, a member of APS, MRS and ECS. He has served as the 3D Packaging Chair of the IMAPS National Technical Committee since 2006. He received the 2008 IEEE CPMT Exceptional Technical Achievement Award in May 2008 "for his pioneering contributions to and leadership in 3D integration/packaging"

# Invited Talk:
# "Hydrogenated silicon (Si:H) thin film solar cells"
## C.R. Wronski
## Pennsylvania State University

In 2009 the manufacture of thin film solar cell panels became a significant fraction of the ~7000 MW production of terrestrial solar PV. This includes the production of what are commonly known as amorphous silicon solar cells. In reality such cells have evolved into what **is** now a hydrogenated silicon thin film technology which last year has seen significant expansion of existing production facilities as well as introduction of numerous turn key factories. After briefly discussing the importance of terrestrial photovoltaics and the currently dominant crystalline silicon technology, the talk reviews a number of issues related to hydrogenated silicon thin film technology. This includes the original innovations in hydrogenated amorphous silicon solar cells as well as the evolution into the improved materials and variety of different solar cell structures that make up the current status of this technology.

## *Curriculum vitae*

In 1987, after 20 years in industry, Dr Wronski joined Pennsylvania State University as Leonhard Professor where he is currently Professor Emeritus. Dr.Wronksi's contributions to photovoltaics began in 1974, with the pioneering work carried out with David Carlson that led to the invention and development of thin film amorphous silicon solar cells. In 1976, he discovered with David Staebler the reversible, light-induced changes in the optoelectronic properties in amorphous silicon known as the Staebler-Wronksi Effect which is still of great scientific and technological interest. He later collaborated in, establishing the limitations imposed by optical absorption on short circuit currents and then demonstrating and quantifying the large improvements possible from the optical enhancement obtained with texturized surfaces. This technique was and is still being applied and refined in all the thin film solar cell technologies. At Penn State the research carried out with Professor Rob Collins for the first time characterized in real-time the growth and microstructure hydrogenated silicon films and cells. The understanding and control of the more ordered structure in "protocrystalline" materials has led to the more systematic tailoring of solar cells which has been utilized by industry. In 1984 Professor Wronksi received together with David Carlson the IEEE Morris N. Liebmann Memorial Award for "crucial contributions to the use of amorphous silicon in low cost, high performance photovoltaic solar cells." In 2000 he received the IEEE W Cherry Award for "Outstanding contributions to photovoltaic science and technology". Professor Wronski has 9 issued patents and over 350 publications on amorphous materials and their devices.

978-1-4244-6572-9/10 $26.00 © 2010 IEEE

# Invited Tutorial:
# "Carbon based Nanomaterials as Interconnects and Passives for Next-Generation VLSI and 3-D ICs"

Kaustav Banerjee

Electrical and Computer Engineering Department

University of California, Santa Barbara

kaustav@ece.ucsb.edu

As IC feature sizes continue to be scaled below 45 nanometer, copper wires exhibit significant "size effects" resulting in a sharp rise in their resistivity, which, in turn, has adverse impact both on their performance as well as reliability---in the form of current carrying capacity. This limitation of copper interconnects has been highlighted by leading semiconductor companies around the world as well as in the International Technology Roadmap for Semiconductors (ITRS), and threatens to slow down or even stall the traditional growth of the semiconductor and related industries.

Carbon based nanomaterials: carbon nanotubes (single, double or multi-walled) and graphene nano-ribbons are known to have amazing electrical, thermal and mechanical properties, and can potentially address the challenges faced by copper and thereby extend the lifetime of "electrical interconnects". Most of these outstanding properties arise from their ultra-strong $sp^2$ hybridized bonds and "low-dimensionality"---they are essentially 1-dimensional structures. This talk will provide a brief overview of these nanomaterials and discuss their prospects as interconnects and passives and how they can be efficiently integrated into the VLSI process to address the dire need for faster, more energy-efficient and more reliable on-chip wiring, and also open new opportunities in integrated energy-storage as well as in off-chip wiring and packaging and in emerging 3-D integrated circuits.

**Keywords**

Carbon nanomaterials, Carbon nanotubes, Graphene, Interconnects, Passives, 3-D ICs, Through-Si Vias.

**References**

[1] H. Li, C. Xu, N. Srivastava and K. Banerjee, "Carbon Nanomaterials for Next Generation Interconnects and Passives: Physics, Status and Prospects," *IEEE Transactions on Electron Devices*, Vol. 56, No. 9, pp. 1799-1821, 2009.

[2] H. Li and K. Banerjee, "High-Frequency Analysis of Carbon Nanotube Interconnects and Implications for On-Chip Inductor Design," *IEEE Transactions on Electron Devices*, Vol. 56, No. 10, pp. 2202-2214, Oct. 2009.

[3] C. Xu, H. Li, R. Suaya, K. Banerjee, "Compact AC Modeling and Analysis of Cu, W, and CNT based Through-Silicon Vias (TSVs) in 3-D ICs," *IEEE International Electron Devices Meeting*, pp. 521-524, 2009.

[4] H. Li, C. Xu and K. Banerjee, "Carbon Nanomaterials: The Ideal Interconnect Technology for Next-Generation ICs," *IEEE Design and Test of Computers, Special Issue on "Emerging Interconnect Technologies for Gigascale Integration"* 2010 (*to appear*).

*Curriculum vitae*

Kaustav Banerjee is a professor in the Electrical and Computer Engineering Department at UC Santa Barbara. He received his Ph.D. degree in electrical engineering and computer sciences from UC Berkeley in 1999. Before joining UCSB in 2002, he was a Research Associate at the Center for Integrated Systems in Stanford University and a visiting faculty at Intel's Circuit Research Lab in Hillsboro, OR. Prof. Banerjee's research interests are in the area of nanoscale VLSI and emerging nanoelectronics. His research has been chronicled in over 200 papers and 3 book chapters, and recognized with numerous awards and honors. He is a Distinguished Lecturer of the IEEE Electron Devices Society. His website containing more information about his research and links to relevant papers can be found at: http://nrl.ece.ucsb.edu/

# Invited Tutorial:
## "Practical Semiconductor Reliability"
### Todd Marquart
### Micron Technology, Boise Idaho

In general more accurate and realistic approaches to reliability are needed as scaling continues to push the true product lifetime much closer to the required field lifetimes. When lifetime margins were wide, being conservative on an analysis was not a big issue. Now, however, estimates must be more accurate since extending the lifetime needlessly costs time and money, while overestimating the lifetime could result in a significant field risk. In general we can no longer afford unrealistic extrapolations, acceleration models or analysis techniques. The days of a "cookbook" approach to reliability are over.

This tutorial is an introduction to semiconductor reliability with emphasis on practical application of analysis techniques and a basic understanding of typical stresses applied to semiconductors. The tutorial has been kept general making applicable to nearly all semiconductor products. This is a subset of a 16 hour course currently taught at Micron by the author. An outline of the topics to be covered is shown below. The special topics listed at the end will be covered if time permits.

RELIABILITY STATISTICS
- *Introduction to the basic statistical tools for life data analysis*
- *Emphasis is on Weibull analysis, its applications and pitfalls*

ACCELERATED TESTING
- *Theory of acceleration factors*
- *Typical acceleration factor models*

BASICS OF INTRINSIC RELIABILITY
- *Electromigration*
- *Stress Voiding*
- *Hot Carrier Injection*
- *Time Dependent Dielectric Breakdown*

BASICS OF ENVIRONMENTAL STRESSING
- *Operational Lifetime Stressing*
- *Corrosion Testing*
- *Thermal-Mechanical Testing*
- *ESD/Latch-up*

SPECIAL TOPICS
- *Introduction to System Reliability*
- *Introduction to Monte-Carlo Simulation*

## *Curriculum vitae*

Todd Marquart received a B.S. in chemistry from Rider College and a Ph.D. in inorganic chemistry from the University of Illinois, Urbana-Champaign. During his chemistry career he worked on a variety of subjects including superconductors, zeolites, low dimensional conductors, phase-transitions and numerical simulation. He did a post-doc at Sandia National labs where he worked in the organic materials group studying nanotubes and fullerenes. He then joined Motorola Semiconductor Products Sector working on the reliability of ASICs, DSPs, NOR flash and MRAM eventually becoming a Distinguished Member of the Technical Staff. After 8 years he joined Micron Technologies where, for the past 6 years, he has primarily, been working on the reliability of NAND flash and SSDs. Todd is currently a Distinguished Member of the Technical Staff in Micron's process R&D. During his time at Motorola he developed and taught the class "Practical Semiconductor Reliability Engineering" and has developed and is teaching a similar class at Micron. Throughout his reliability career he has been a contributor to the "New Weibull Handbook" as well as the Weibull software package "WinSMITH Weibull". Todd's emphasis has been on developing practical, best-practice approaches to life-data analysis which is required to deal with the real, non-ideal data encountered regularly in engineering.

978-1-4244-6572-9/10 $26.00 © 2010 IEEE

# TECHNICAL PRESENTATIONS

978-1-4244-6572-9/10 $26.00 © 2010 IEEE

# Modified Floating Gate and IPD Profile for Better Cell Performance of Sub-50 nm NAND Flash Memory

Jennifer Lequn Liu, Fernando Gonzalez, and Y. Jeff Hu
R&D Process Development
Micron Technology, Inc.
Boise, ID-83707, U.S.A

Jixin Yu, Charan Srinivasan, Ervin Hill
Intel JDP
Boise, ID-83707, U.S.A

*Abstract—* We report a new approach to utilize oxygen implantation on the top of the floating gate (FG) to improve the cell performance of a sub-50 nm NAND flash memory cell. This method was used to form a thin oxide layer only on the top of the FG but not on the sidewalls. It also rounded the corners of the FG. As a result, the leakage current between FG and control gate (CG) was reduced without sacrificing the gate coupling ratio (GCR). With this approach we improved $V_{t\_sat\_program}$ and $V_{t\_sat\_erase}$ without degrading $V_gV_{t\_program}$ and $V_gV_{t\_erase}$ on real Si.

*Keywords-oxygen implant; floating gate; corner rounding; NAND*

## I. INTRODUCTION

There are several challenges to continue scaling the sub-50 nm floating gate (FG) NAND cell while maintaining good data retention [1-3]. With shrinking dimensions, the leakage through the inter-poly dielectric (IPD) has increased. One of the key issues is how to reduce the leakage current between FG and control gate (CG) while not sacrificing the gate coupling ratio (GCR) at the same time. Based on the TCAD simulation, leakage current happens mostly at the top and sharp corners of the FG due to the high electric field at those locations. If we can reduce the electric field, we can reduce the leakage current between FG and CG, and consequently improve the cell reliability.

In this paper we report a new approach to reduce the electric field by implanting oxygen directly into the top of the poly FG. This method reduced the electric field at those locations and improved the cell performance.

## II. EXPERIMENTAL

Oxygen was implanted with low energy but high dose directly on the top of FG poly. This resulted in a high concentration of oxygen close to the poly FG surface after implant. The implant condition was based on TRIM simulation results. In this experiment we focused on an implant energy of 1KeV and a dose ranging from 1E15 to 1E16 cm$^{-2}$.

During the oxygen implant oxygen species bombarded the FG surface, rounding the sharp corners of the FG. Implants with and without tilt angle were both explored on the device wafers for better corner rounding. The tilt angle range tested was zero to 45 degrees.

Wafers were then annealed in an $N_2$ environment at high temperature. During annealing, implanted oxygen atoms reacted with silicon atoms in the implanted area, and formed a thin oxide layer on top of the FG poly. The thermal cycles of IPD growth could also contribute to this oxidation. Because there was no oxygen implanted into the sidewalls, the thin oxide layer was only formed on the top and corners of the FG poly but not on the sidewalls.

The bottom oxide of the IPD film stack was then grown by a single wafer radical oxidation (RADOX) process. RADOX growth temperatures of 850°C and 950°C were compared. RADOX thickness was kept constant. Since there was already a thin oxide layer on the top and corners of the FG poly, the total oxide on the top and corners of FG poly was thicker than it would be without the oxygen implant. The sidewall oxide thickness was unchanged between the two cases.

## III. RESULTS AND DISCUSSION

We studied the oxide growth rate using blank poly Si wafers. The wafers were implanted with oxygen and then

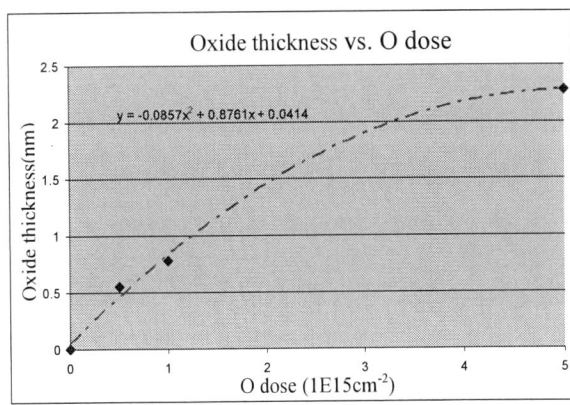

Figure 1. Oxide thickness formed on the top of poly after 15 second anneal at 950°C vs. oxygen implant dose. Oxygen implant energy was 1KeV. Oxide thickness was measured by ARXPS.

978-1-4244-6572-9/10 $26.00 © 2010 IEEE

annealed at 950°C for 15 seconds. The correlation between oxygen implant dose and oxide thickness is shown in Fig. 1. The oxide thickness was measured by ARXPS. Oxide thickness generally increased with higher oxygen dose. An 850 °C thermal anneal was also tested. There was no significant difference in oxide growth rate between 850°C and 950°C.

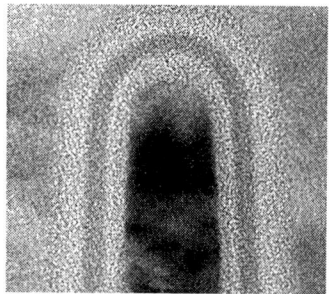

Figure 2. FG poly and IPD profile modified by oxygen implant and post implant anneal. Implant energy is 1KeV, dose is 1E16cm$^{-2}$. Post implant anneal was done at 950°C for 15 second.

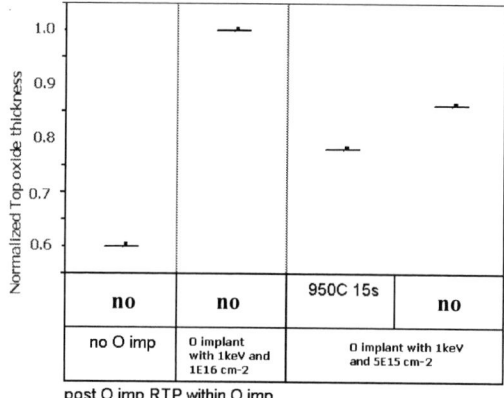

Figure 3. top oxide thickness showed: O implant introduces extra oxide on the top of poly, and total top oxide layer is thicker than that without imp. Higher O dose caused thicker top oxide. Post implant anneal has little impact on top oxide thickness.

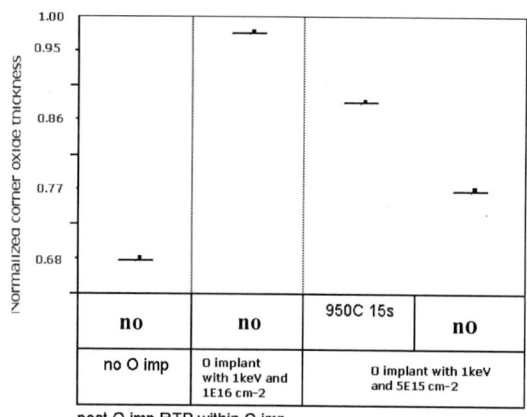

Figure 4. corner oxide thickness showed: O implant introduces extra oxide at corner of poly, and total oxide layer is thicker than that without imp. Higher O dose gets thicker oxide. Post anneal has little impact on corner oxide thickness.

The IPD bottom oxide thickness at the top, corners and sidewalls of the FG, as well as the IPD/FG poly corner curvature were studied on the structure wafers by TEM. Fig. 2 is an example of a TEM picture showing the FG poly and IPD profile modified by Oxygen implant. The FG poly was implanted with 1E16 cm$^{-2}$ oxygen at 1KeV. Before the IPD bottom oxide growth the wafers were annealed at 950°C for 15 second to form oxide on the implanted area. After IPD growth the total oxide on the top of the FG poly was ~70% thicker than that on the sidewalls. In addition the total oxide on the corner of the FG poly was ~30% thicker than that on the sidewalls. Though not shown here, the FG poly corner was also rounded further compared to the profile without implant.

We also tested different oxygen implant doses from 1E15 cm$^{-2}$ to 1E16 cm$^{-2}$, with and without the extra post implant anneal (Fig. 3-6). Top/corner/sidewall oxide thicknesses and IPD/poly FG corner curvature were measured by TEM. It showed that top/corner oxide thickness was increased with higher implant dose, but sidewall oxide thickness remained same as that without implant. The FG poly corner curvature increased with higher implant dose as well. The post implant

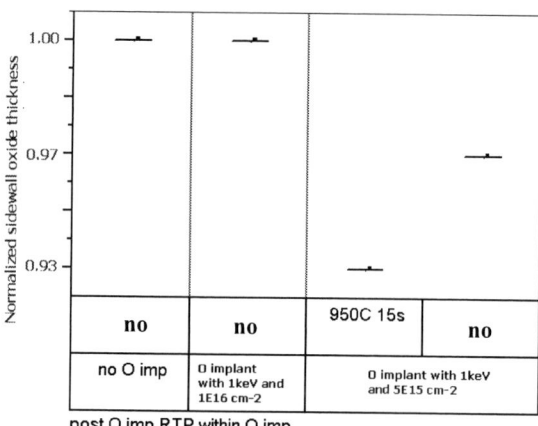

Figure 5. sidewall oxide thickness showed Oxygen implant and post anneal do not make significant difference on sidewall.

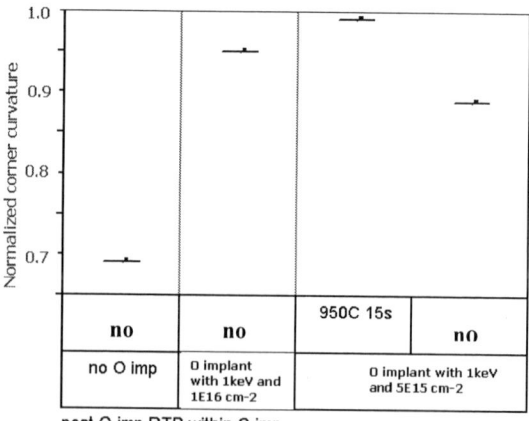

Figure 6. corner curvature showed: Oxygen implant made corner rounded more than without implant. Post implant anneal has no significant impact on it.

978-1-4244-6572-9/10 $26.00 © 2010 IEEE

anneal did not impact oxide thickness or corner curvature significantly. Since there are IPD thermal processes which also help to form oxide after implant, the extra post implant anneal could be skipped to keep the same thermal budget as without implant.

C-V (Fig. 7) and I-V (Fig. 8) curves were measured on field edge test structures. The oxygen implant was then done with 1KeV energy and 5E15 cm$^{-2}$ to 1E16 cm$^{-2}$ dose range. The post implant anneal was done at 950°C for 15 seconds. The C-V curve showed that the oxygen implant increased the electrical oxide thickness (EOT) of the IPD. This is consistent with thicker oxide at the FG top and corners. The EOT also increased with higher oxygen implant dose. The post anneal did not impact EOT significantly. The I-V curve showed that

the oxygen implant reduced leakage current. The higher oxygen dose caused less leakage current. The post implant anneal also did not significantly impact leakage current. The C-V and I-V data were in agreement with the TEM results.

It should be pointed out that from the C-V curve we could also see that the FG poly depletion was increased by oxygen implantation. Dopant was lost during the oxidation process which consumes Si, thus reducing the total dopant concentration in the FG poly. As we will discuss further in the following paragraphs, depletion may have a negative impact on cell performance and may reduce the benefits from the IPD/FG profile improvement by oxygen implant.

$V_{t\_sat}$ was measured on device wafers with cell structures (Fig. 9). With the 850°C RADOX process, both $V_{t\_sat\_program}$ and $V_{t\_sat\_erase}$ were improved by oxygen implant. The improvement was dominated by the oxygen implant dose. Higher oxygen implant doses showed more $V_{t\_sat}$ improvement. This was consistent with the TEM observation that the IPD bottom oxide thickness on the top and corners of FG poly was increased by oxygen implant. For a 1E16 cm$^{-2}$ dose, $V_{t\_sat\_program}$ was improved by about 5%, and $V_{t\_sat\_erase}$ was improved by 2~3%.

Implant tilt angle did not have a significant impact on $V_{t\_sat\_program}$. On the other hand, implanting with a higher tilt angle resulted in a significant improvement on $V_{t\_sat\_erase}$. The angled implant could round the corner more than a zero angle implant, reducing the corner electrical field further. This indicates that the corner electrical field may have more impact on $V_{t\_sat}$ during erase than during program.

However, we did not observe $V_{t\_sat}$ improvement with the

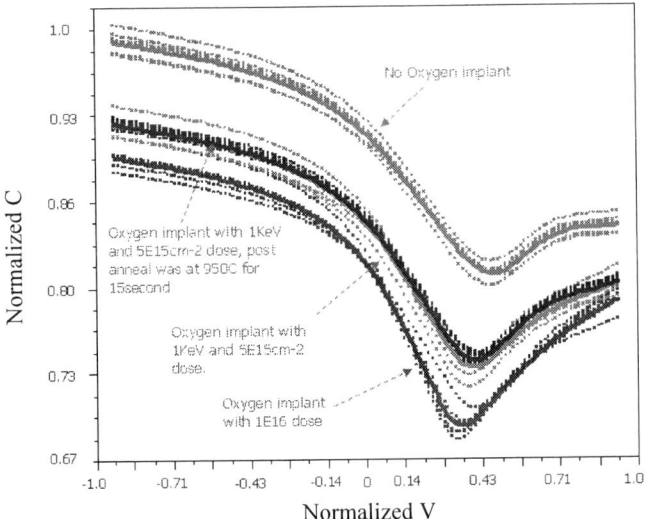

Figure 7. C-V curve showed electrical IPD film stack thickness was increased with higher implant dose. Post oxygen implant anneal makes the curve tighter with less variation.

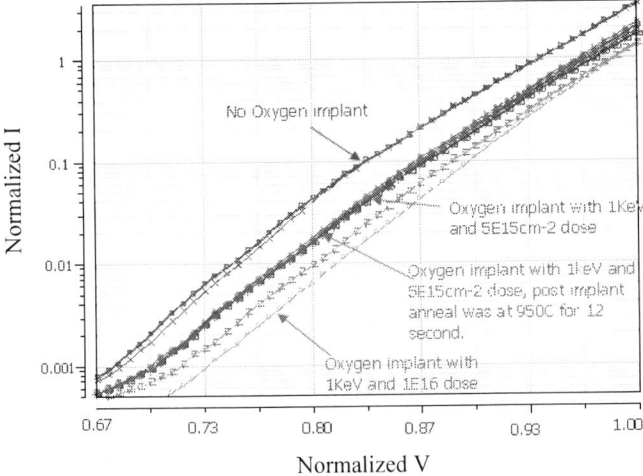

Figure 8. I-V curve showed leakage current was improved by higher oxygen implant dose. Post oxygen implant anneal does not impact leakage current much.

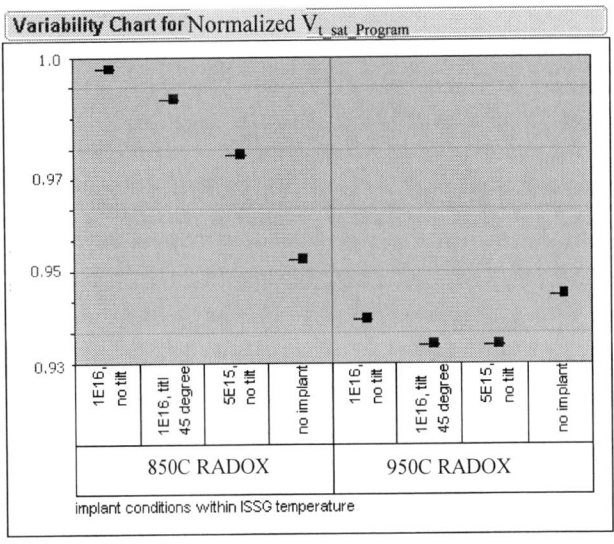

Figure 9. Oxygen implant was done with 1KeV for all groups. $V_{t\_sat\_program}$ was mainly increased with higher oxygen dose. There was up to ~5% improvement in $V_{t\_sat\_program}$ with 1E16 dose and 850°C RADOX, but not much improvement with 950°C RADOX possibly due to more dopant loss. Tilt angle had an insignificant impact on $V_{t\_sat\_program}$.

950 °C RADOX process. Since our TEM showed no significant difference in IPD/FG poly profiles between 850°C and 950°C RADOX processes, we believe the difference between the 850°C and 950°C processes may result from more dopant out-diffusion at 950°C than at 850°C. Based on our data, oxygen implantation increases the IPD thickness at the top and corner of the FG poly, and should increase $V_{t\_sat}$. At the same time, the high temperature RADOX process also causes dopant to out-diffuse into the IPD bottom oxide from the FG poly. This reduced the dopant density in the FG poly and thus increased FG poly depletion. The 850°C and 950°C C-V curves also showed that the 950°C process caused more poly depletion than the 850 °C process. Since FG poly depletion is known to reduce $V_{t\_sat}$, more dopant loss will reduce $V_{t\_sat}$ more. Other factors impacting depletion are increased oxygen implant dose and FG oxidation during anneal. Since dopant loss competes with the improvements resulting from the improved IPD/FG profile by oxygen implant, the final $V_{t\_sat}$ value will depend upon which mechanism prevails. It is critical to optimize the oxygen implant conditions and RADOX temperature, and to make sure the $V_{t\_sat}$ improvement is maintained. Based on our data with oxygen implant energy at 1KeV and dose between 5E15 cm$^{-2}$ to 1E16 cm$^{-2}$, dopant loss is the dominant effect at 950°C, while the IPD/FG profile improvement by oxygen implant is the dominant effect at 850°C. We may need to either limit the RADOX temperature to 850°C or below, or increase dopant density in the FG poly in order to maintain $V_{t\_sat}$ improvement. Simultaneously optimizing the implant dose may be necessary to gain maximum $V_{t\_sat}$ improvement.

$V_g V_{t\_program}$ and $V_g V_{t\_erase}$ were also measured on device wafers. The oxygen implant did not show significant impact on them, indicating that the GCR was not degraded by the oxygen implant.

## IV. SUMMARY

The FG poly and IPD profile were modified by oxygen implantation in order to reduce the electric field across the IPD. Consequently the leakage current between the FG and the CG was reduced. We successfully thickened the IPD bottom oxide layer on the top and corners of the FG poly, while keeping the same oxide thickness at the FG sidewalls. We also rounded the FG poly corner further by oxygen implant. We discussed two competing mechanisms which impact cell performance: dopant loss which negatively impacted performance and IPD/FG profile improvement by oxygen implant which showed positive performance benefits. We provided direction for engineering the growth temperature of the IPD bottom oxide, the oxygen implant dose, and the poly FG dopant density to achieve optimal results. As a result,

we achieved better $V_{t\_sat\_program}$ and $V_{t\_sat\_erase}$ without degrading $V_g V_{t\_program}$ or $V_g V_{t\_erase}$. This approach can be used as a tool to improve the cell performance of sub-50 nm FG NAND.

## ACKNOWLEDGEMENT

The authors would like to thank the Micron surface lab and TEM lab for material characterization, Karl Holtzclaw for device measurements, and Haitao Liu, Shu Qin and Allen Mcteer for helpful discussions.

## REFERENCE

[1] K.N.Kim, et al., "The future prospect of non-volatile memory" Technical Digest VSLI-TSA, pp.88-94, 2005.
[2] Y.Shin et al., "Non-volatile memory technologies for beyond 2010" Technical Digest VLSI, 2005.
[3] K.Kim, "Technology for sub-50nm DRAM and NAND flash manufacturing" IEDM Technical Digest, December 5-7, 2005, pp. 323-326.

# Study of Carrier Mobility of Low-Energy, High-Dose Ion Implantations using Continuous Anodic Oxidation Technique/Differential Hall Effect (CAOT/DHE) Measurements

Shu Qin, Y. Jeff Hu, and Allen McTeer

Process R/D Department
Micron Technology, Inc.
Boise, ID, U.S.A
sqin@micron.com

Si Prussin and Jason Reyes

Department of Electrical Engineering
University of California at Los Angeles
Los Angeles, CA, U.S.A

*Abstract*—New carrier mobility ($\mu$) data for boron-, phosphorus-, and arsenic-doped Si in a low-energy, high-dose implant regime are measured and studied using continuous anodic oxidation technique/differential Hall effect (CAOT/DHE) technique. The data show that when the doping concentration is $>10^{20}/cm^3$, both hole and electron mobility values are lower than the conventional model predictions, and the electron mobility of the As-doped Si is lower than the P-doped ones. The data also show that when the doping concentration is $>10^{21}/cm^3$, the hole mobility in B-doped Si and the electron mobility in P-doped Si are almost equal and reach as low as ~40 $cm^2/V$ sec, and the electron mobility of As-doped Si is the lowest and reaches ~30 $cm^2/V$ sec. These mobility data are much lower than the conventional model predictions and are also lower than the previously published data. For the ULSI device and circuit analyses, simulations, and designs, these new mobility data need to be taken into consideration.

*Keywords-Carrier and mobility profiles, continuous anodic oxidation technique/differential Hall effect (CAOT/DHE) method, spreading resistance profiling (SRP) method, low energy high dose implants, plasma doping (PLAD).*

## I. INTRODUCTION

The carrier mobility ($\mu$) values of both hole and electron in Si are very important parameters for ULSI circuit and device analysis and design. It is well known that the carrier mobility is dependent upon dopant or carrier concentrations, and the hole mobility is ~one-third of the electron mobility at the low and moderate doping level. The carrier mobility decreases with increasing impurity or carrier concentrations due to impurity and/or carrier scattering mechanisms. This is a very important guideline, or rule, for circuit and device analyses and layout designs. However, the information in the literature about the mobility scattering and degradation mechanisms can be confusing and misleading. Some literature lists the mobility versus the impurity concentration [1–5], and some literature lists the mobility versus the carrier concentration [5–7].

Two aspects of the information presented may cause confusion. First is the doping and activation regime. In the low doping level and full activation regime, such as thermal diffusion, because impurities are fully (100%) ionized, the carrier concentration is equal to the impurity concentration. However, for ULSI device processing, low- or ultra-low–energy, high-dose ion implantation and rapid thermal process

(RTP) annealing are being used for doping and activation to meet a requirement of ultra-shallow junction depth ($x_j$), suitable junction abruptness, and lower sheet resistance ($R_S$). The impurity concentration can be higher than the impurity solid solubility. Because the impurity solid solubility is dependent on RTP process temperature, which is limited for $x_j$ control, the carrier concentrations are limited by the impurity solid solubility, which is ~$1.8\times10^{20}/cm^3$ for boron, ~$1.0\times10^{21}/cm^3$ for phosphorus, and ~$1.8\times10^{21}/cm^3$ for arsenic, when activation temperature is 1000°C [8]. Therefore, the impurity concentration can be much higher than the carrier concentration in this process regime.

The old mobility data were based on the thermal process, such as diffusion, in which the impurity concentrations were lower than solid solubility, and dopants were fully activated and annealed. In the literature, there is a lack of mobility data for low- or ultra-low-energy, high-dose ion implanted Si wafers, where impurity concentrations are $>10^{20}/cm^3$ and are with more lattice defects and damages. Implant damages can adversely affect mobility. For ULSI devices, to retard the short channel effect (SCE), the channel background dopant level had been increased from $10^{18}$ to the $10^{19}/cm^3$ range, and the dopant level of the source and drain extension (SDE) regions can be much higher than $10^{20}/cm^3$. Therefore, as the device size is scaled down, the device speed is limited more and more by the carrier mobility in the higher dopant level regime.

The second aspect of the information that may be misleading is the metrology limits. So far, the spreading resistance profiling (SRP) method is widely used as a commercial service for carrier concentration profiling measurements [9,10]. However, this method has some fundamental issues for measurements of higher dopant concentrations $>1\times10^{20}/cm^3$ because it does not take into account the effect of defects or damage caused by ion implantation and advanced annealing techniques, such as laser- or flash-based anneals [11]. The conventional differential Hall effect (DHE) method has been used for carrier and mobility profiling measurements [12]. However, this method has issues for commercialization due to the complexity of the procedure and measurement accuracy, making it time- and cost-prohibitive [11].

Secondary ion mass spectrometry (SIMS) is another technique in the semiconductor processing community that is widely used to measure and monitor the impurity profile shape and dose of the implanted wafers. Because SIMS is a relatively cheaper, more convenient, and standard method for measuring an accurate impurity profile in Si, from an engineering point of view, the study of the mobility as a function of the impurity concentration is more useful, less debatable, and more processing-oriented for the device and circuit analyses, simulations, and designs.

In this study, continuous anodic oxidation technique/differential Hall effect (CAOT/DHE) method is used to measure the carrier mobility of both hole and electron as a function of the impurity concentrations of the low-energy, high-dose ion implantations [11,13]. A newly-developed surface analysis technique, secondary ion mass spectrometry/angle-resolved x-ray photoelectron spectroscopy (SIMS/ARXPS) method, is utilized to characterize impurity profiles and doses [14]. Because it can measure and decouple the native oxide, the impurity profiles and doses can be more accurate. The ion implants include boron, phosphorus, and arsenic conventional beam-line implants and $B_2H_6$ and $AsH_3$ plasma doping (PLAD). The implanted wafers were annealed using an RTP to activate the impurities with a condition of 995°C and 20 sec in $N_2$ ambient.

## II. EXPERIMENTS AND DISCUSSIONS

Fig. 1 shows a comparison of the retained B profiles of SIMS/ARXPS [14] and the carrier (hole) and mobility ($\mu$) profiles of CAOT/DHE [11] of (A) the conventional beam-line B ion implant with an energy of 2keV and a dose of $5\times10^{15}$/cm$^2$, and (B) $B_2H_6$ PLAD with a voltage of -6kV and a total dose of $2\times10^{16}$/cm$^2$. The values next to the curves in Fig. 1 are SIMS-measured, B-retained doses in Si, which have extracted the B loss in the surface oxides, and carrier (hole) doses and averaged mobility measured by CAOT/DHE, respectively.

For both beam-line implant and PLAD, several interesting results are shown in Fig. 1. First of all, both implanted wafers show very high B concentrations near the surface, which are in the $10^{21}$ to $10^{22}$/cm$^3$ range and much higher than the B solid solubility ($B_{SS}$) of the current RTP temperature. Second, the CAOT/DHE method offers carrier and carrier drift mobility profiles, and demonstrates that near the Si surface, the mobility decreases with increasing B concentration and is much lower than an average mobility due to carrier scattering mechanisms with heavily doped impurities as well as with more serious lattice defects and damages. Third, the carrier profiles of CAOT/DHE measurement are very consistent with B profiles of SIMS measurement beneath $B_{SS}$ levels except for those near the surface. This amazing feature has never been shown by the conventional SRP or DHE methods previously and demonstrated that the CAOT/DHE method is a powerful method for studying doping and activation processes. These results experimentally confirm that the implanted B impurities in Si beneath $B_{SS}$ can be fully (~100%) activated under the current RTP condition.

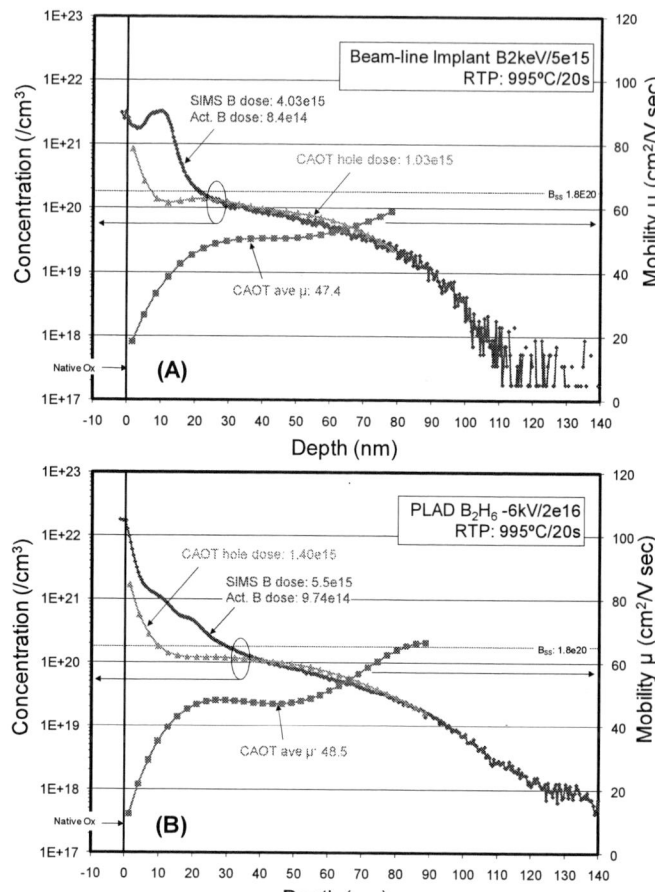

Figure 1. SIMS/ARXPS B profile and CAOT/DHE carrier and $\mu$ profiles of (A) beam-line B 2 keV/$5\times10^{15}$/cm$^2$ implant, and (B) $B_2H_6$ -6kV/$2\times10^{16}$/cm$^2$ PLAD.

Based on the impurity and mobility profiling data, the mobility ($\mu$) of holes is plotted as a function of B impurity concentrations in Fig. 2. For comparison, a conventional empirical Fermi-distribution modeling curve of the hole mobility as a function of B concentration is also plotted in Fig. 2 [1,2]. There is a good consistency and overlap between the modeling predicted data and the measurement data when B concentrations are between $10^{19}$ and $10^{20}$/cm$^3$. When B concentrations are $>10^{20}$/cm$^3$, the modeling predicts that the hole mobility saturates at 47.7 cm$^2$/V sec. However, the measured hole mobility does not saturate and decreases further with B concentrations to as low as ~20–38 cm$^2$/V sec when B concentrations are ~$2\times10^{21}$/cm$^3$. It is noted that for a PLAD B-doped Si wafer, the B concentration can increase to as high as $10^{22}$/cm$^3$, and the hole mobility is reduced further to ~10 cm$^2$/V sec. This is because B concentration peak of PLAD is at the Si surface, and the mobility is degraded further by the extra surface scattering mechanism.

Fig. 3 shows a comparison of the retained n-type impurity profiles of SIMS/ARXPS and the carrier (electron) and mobility profiles of CAOT/DHE of (A) the beam-line P ion implant with an energy of 2keV and a dose of $5\times10^{15}$/cm$^2$, and (B) the beam-line As ion implant with an energy of 10keV and a dose of $5\times10^{15}$/cm$^2$.

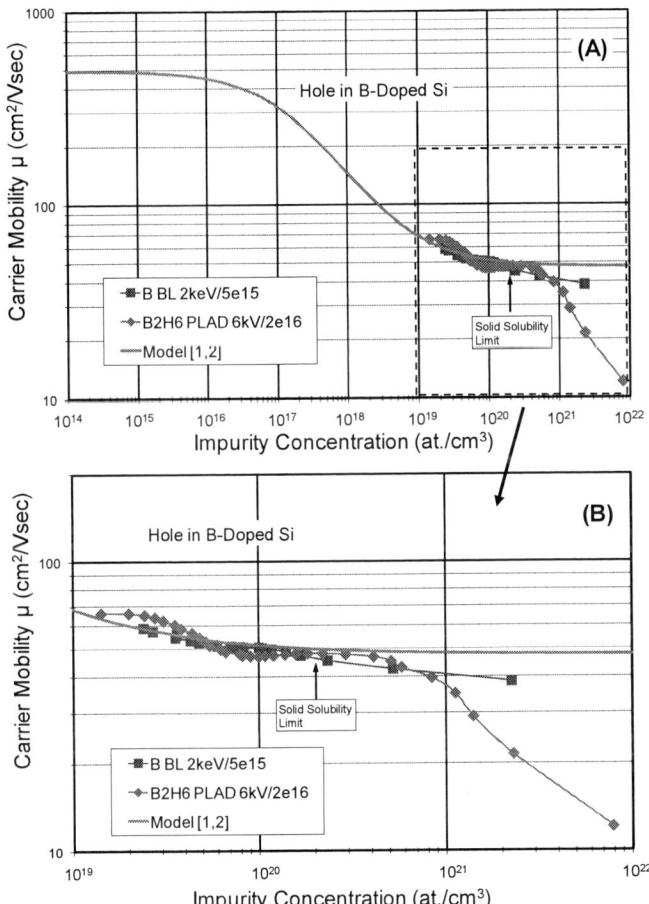

Figure 2. Hole drift mobility versus p-type impurity (B-doped) concentration in Si.

For n-type impurity implants and current RTP condition, both P and As impurity profiles are beneath their solid solubility levels. However, phosphorus-doped and arsenic-doped and boron-doped Si wafers show very different activation behaviors. Of interest is the fraction of dopant dissolved. This is the total carrier dose divided by the total SIMS impurity dose. For beam-line B, it is 25.6%; for PLAD B, it is 25.5%; for beam-line P, it is 68.8%; and for beam-line As, it is 33.8%. More fairly, we can define an activation fraction as the total carrier dose divided by the SIMS activated impurity dose. The activated impurity dose is defined and calculated by integrating the SIMS impurity profile beneath solid solubility. The P-doped implant shows an activation fraction of ~68.8%, even though it is lower than the B-doped case of ~100%. As-doped implant shows a very low activation fraction of ~33.8%, which is only ~1/3 of B-doped one and ~1/2 of P-doped one.

Two main reasons cause such a low activation for As implant. The first reason is because intrinsically the activation energy of As ions is much higher than what is required for B and P ions [15]. Another reason is that an As ion has a much larger atomic mass unit (AMU), which is 75 compared with 31 for P and 11 for B, and requires a higher implant energy than P and B for a comparable $x_j$-$R_S$ characteristics. Therefore, an As implant will involve more lattice defects and damages. These defects and damages will degrade electron mobility and need a higher thermal budget (DT) to anneal. This mobility degradation of As implant can be confirmed in Fig. 3, in which the average mobility of As implant is 49.8 cm²/V sec and lower than 54.5 cm²/V sec of P implant one. Similar to B-doped Si, the CAOT/DHE method offers carrier drift mobility profiles and demonstrates that near the Si surface, the values of mobility decrease with increasing impurity concentrations and are much lower than the average mobility due to carrier scattering mechanisms with heavily doped impurities as well as with more serious lattice defects and damages.

Generally, no distinction was made in the past between the electron mobility in P-doped and in As-doped Si; thus, in recent years, only limited data have been presented for As-doped Si.

Based on the impurity and mobility profiling data, electron mobility is plotted as a function of impurity concentrations in Fig. 4. For comparison, a conventional empirical Fermi-distribution modeling curve of the electron mobility as a function of P concentration is also plotted in Fig. 4 [1,3]. When P concentrations are >$10^{20}$/cm³, the modeling predicts that the electron mobility saturates at 92 cm²/V sec. However, the measured electron mobility does not saturate and decreases further with P concentrations to as low as ~40 cm²/V sec when the P concentration is ~2×$10^{21}$/cm³.

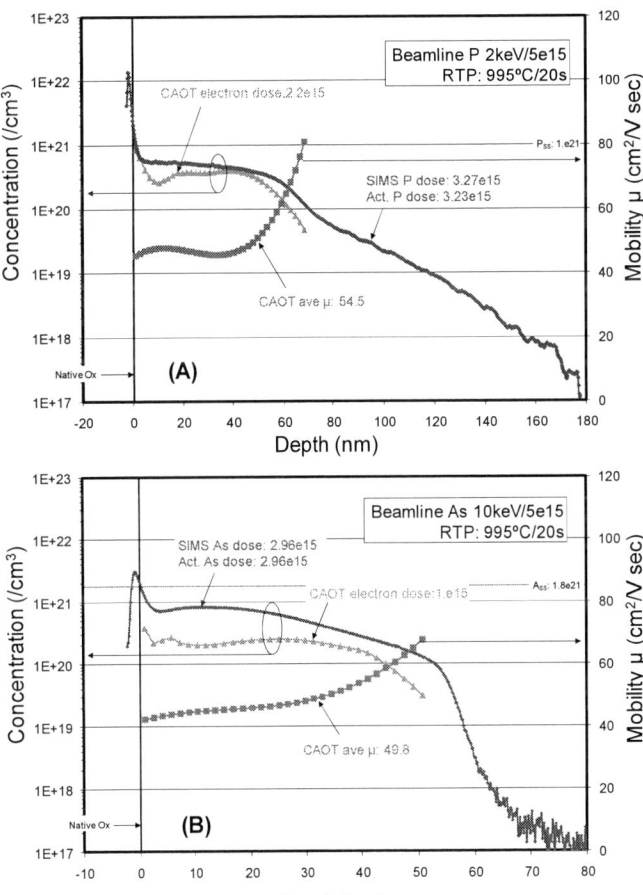

Figure 3. SIMS/ARXPS profile and CAOT/DHE carrier and μ profiles of (A) P 2keV/5×$10^{15}$/cm² implant, and (B) As 10keV/5×$10^{15}$/cm² implant.

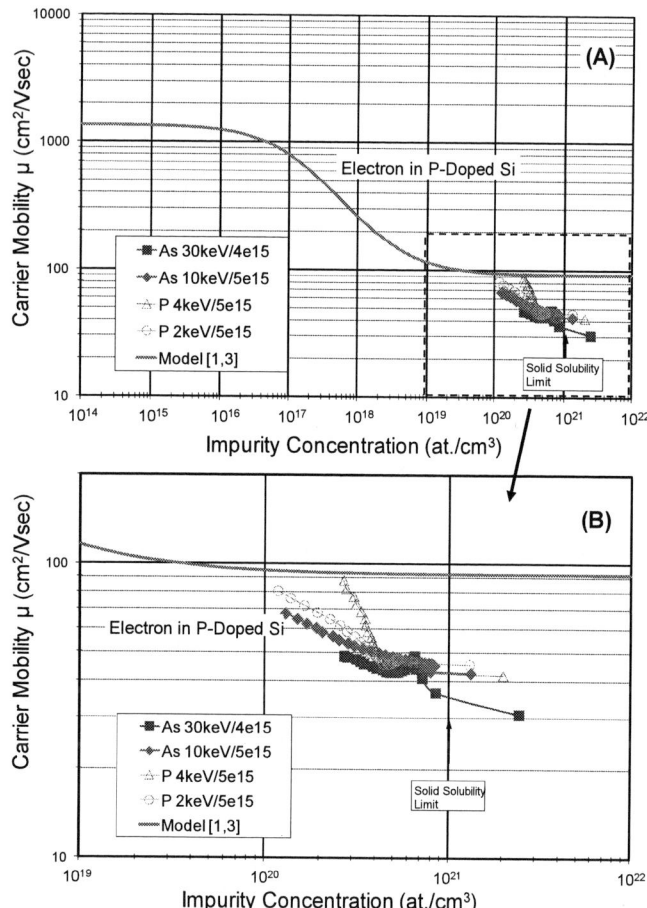

Figure 4. Electron drift mobility versus n-type impurity (P- and As-doped) concentration in Si.

A comparison between the data reported in Fig. 4 indicates that the electron mobility in the As-doped samples is lower than the P-doped ones after the impurity concentrations are $>10^{20}/cm^3$. This result is qualitatively consistent with other experimental data [6]. As mentioned previously, this is because As-doped Si is not fully activated and annealed so that the electron mobility is more degraded. The published data show that the mobility values for the two dopants (P and As) tend to merge for the dopant concentrations below about $10^{19}/cm^3$ [6].

Another interesting feature can be shown if comparing hole and electron mobility. When the impurity concentrations are $10^{20}/cm^3$, the hole mobility is ~47.7 cm²/V sec, and the electron mobility of P-doped and As-doped Si are ~90 and ~70 cm²/V sec, respectively. The electron mobility of P-doped Si and As-doped Si are only ~80% and ~40% higher than hole mobility, respectively. When the impurity concentrations are $>10^{20}/cm^3$, both the hole and electron mobility decrease further with impurity concentrations, but the electron mobility decreases faster. When the impurity concentrations are $2 \times 10^{21}/cm^3$, the electron of P-doped Si and hole of B-doped Si reach a same mobility value of ~40 cm²/V sec, and the electron of As-doped Si reaches the lowest mobility of ~30 cm²/V sec. These new mobility data are much lower than the conventional modeling predicted data [2,3] and also lower than other published experimental data [2-5].

## III. CONCLUSION

New carrier mobility data for boron-, phosphorus-, and arsenic-doped Si in a low-energy, high-dose implant regime are measured and studied using CAOT/DHE technique. The data show that when the doping concentration is $>10^{20}/cm^3$, both hole and electron mobility are lower than the conventional model predictions, and the electron mobility of the As-doped Si is lower than the P-doped ones. The data also show that when the doping concentration is $>10^{21}/cm^3$, both the hole mobility in B-doped Si and the electron mobility in P-doped Si are almost equal and reach as low as ~40 cm²/V sec, and the electron mobility of As-doped Si is the lowest and reaches ~30 cm²/V sec. These mobility values are much lower than the conventional model predictions and also lower than previously published data. For the ULSI device, and circuit analyses, simulations, and designs, these new mobility data need to be taken into consideration.

## REFERENCES

[1] Quick Reference Manual for Semiconductor Engineers, Vol.1, Bell Laboratories, Reading, PA, p. 2-42, December 1982.

[2] S. Wagner, "Diffusion of Boron from Shallow Ion Implants in Silicon", J. Electrochem. Soc.: Solid-State Science and Technology, Vol.119, p. 1570, 1972.

[3] G. Baccarani and P. Ostoja, "Electron Mobility Empirically Related to the Phosphorus Concentration in Silicon", Solid-State Electronics, Vol. 18, p. 579, 1975.

[4] P. Torton, T. Braggins, and H. Levinstein, "Impurity and lattice scattering parameters as determined from Hall and mobility analysis in n-type silicon", Phys. Rev., vol. B8, no. 12, pp.5632-5653, Dec. 1973.

[5] D. K. Schroder, Semiconductor Material and Device Characterization, 2nd Ed., John Wiley & Sons, New York, NY, 1998, pp. 551-553.

[6] G. Masetti, M. Severi, and S. Solmi, "Modeling of Carrier Mobility Against Carrier Concentration in Arsenic-, Phosphorus-, and Boron-Doped Silicon", IEEE Trans. on Electron Devices, vol. ED-30, no. 7, pp. 764-769, July 1983.

[7] N. S. Bennett, N. E. B. Cowern, and B. J. Sealy, "Model for electron mobility as a function of carrier concentration and strain in heavily doped strained silicon", Appl. Phys. Letters, vol. 94, pp. 252109, 2009.

[8] S. K. Ghandhi, VLSI Fabrication Principles Silicon and Gallium Arsenide, John Wiley & Sons, Inc., New York, NY, 1994, pp. 90.

[9] R. G. Mazur and D. H. Dickey, "A Spreading Resistance Technique for Resistivity Measurements in Si", J. Electrochem. Soc., vol. 113, pp. 255-259, 1966.

[10] J. R. Ehrstein, "Spreading Resistance Measurements–An Overview", in Emerging Semiconductor Technology (D. C. Gupta and R. P. Langer, eds.), STP 960, Am. Soc. Test. Mat., Philadelphia, 1987, pp. 453-479.

[11] S. Qin, S. Prussin, J. Reyes, Y. Jeff Hu, A. Mcteer, "Study of Low-Energy Doping Processes using Continuous Anodic Oxidation Technique/Differential Hall Effect (CAOT/DHE) Measurements", IEEE Trans. on Plasma Science, vol. 37, no. 9, pp.1754-1759, September 2009.

[12] N. G. E. Johansson, J. W. Mayer, O. J. Marsh, "Technique used in Hall effect analysis of ion implanted Si and Ge", Solid-State Electronics, vol. 13, no. 3, pp. 317-335, 1970.

[13] S. Prussin, S. Qin, J. Reyes, and A. McTeer, "The Application of the Continuous Anodic Oxidation Technique for the Evaluation of State-of-the-Art Front-End Structures", The 17th International Conference on Ion Implantation Technology (IIT-2008), Monterey, CA, USA, June 08-13, 2008, AIP Conference Proceedings, Vol. 1066, pp. 75-78, 2008.

[14] S. Qin, K. Zhuang, S. Lu, A. McTeer, W. Morinville, and K. Noehring, "SIMS/ARXPS – A New Technique of Retained Dopant Dose and Profile Measurement of Ultra-Low Energy Doping Processes", IEEE Trans. on Plasma Science, vol. 37, no. 1, pp. 139-145, January 2009.

[15] S. K. Ghandhi, VLSI Fabrication Principles Silicon and Gallium Arsenide, John Wiley & Sons, Inc., New York, NY, 1994, p. 247.

978-1-4244-6572-9/10 $26.00 © 2010 IEEE

# Discrete Test Structure Device Degradation Analysis and Correlation to NAND Flash Circuit Operation

Jasper Gibbons, Puneet Sharma, Steve Porter, Jim Fulford, Praveen Vaidyanathan, Sheryll De Guzman, Pratap Murali, Ken Marr

Micron Technology, Inc.
Research and Development
Boise, ID, United States

*Abstract*- **A methodology is established to correlate shifts of test structure device parameters, due to device degradation or process variation, to circuit operation throughout the product lifetime. To the authors' knowledge, this work is original in that SPICE simulation is used, with degraded device models, to relate a circuit timing metric to the degradation of a discrete device used in the circuit. The correlation is validated with actual circuit measurements. In this study, the NAND Flash high-voltage switch circuit is examined in regards to the effect of degrading the p-channel MOSFET used in the circuit. Under standard operating conditions, the device degrades under high electric fields applied across the gate oxide. The method enables the accurate prediction of product lifetime using test structure measurements.**

*Keywords: NAND, high voltage; SPICE; test structure; HV; Switch; FN tunneling; trapped charge; PMOS; flash; MLC; rise time; time delay; level shifter*

## I. Introduction

With the increasingly competitive market for higher density NAND Flash memory [1], companies have the need to aggressively scale devices to smaller dimensions [2]. The voltages required for program and erase operations continue to increase with scaling and the demand for MLC (multi-level-cell) functionality [3-5]. Ensuring the reliability of high-voltage transistors is an important challenge for memory manufacturers. On latest generation MLC parts, nodes can be biased up to 29V, which results in an electric field of 8.3MV/cm across a typical high-voltage gate oxide of 350A. Under this condition, Fowler-Nordheim tunneling occurs. The tunneling results in charge trapping in the gate oxide and the creation of interface states [6-7]. The trapped charge changes the threshold voltage ($V_t$) of the device and can result in critical circuits not functioning properly. Also, the variation of device $V_t$ due to process variation in manufacturing will impact circuit functionality.

The primary purpose of this study is to develop, and verify, a method for predicting the lifetime of a critical circuit used in NAND Flash memory products, referred to as the "live die" in this paper, with the results of degradation testing on discrete test structure devices. Therefore, it is necessary to correlate test structure data to degraded circuit testing data.

The high-voltage switch circuit is used to fully pass the high voltage from one stage to the subsequent stage without a $V_t$ drop. It is placed at each stage of the decode path and used to pass the high voltage pump output to the wordline or substrate of the page or block that needs to be programmed or erased, respectively [8]. Therefore, if the circuit fails, it can result in the entire die, or portions of, not functioning for basic operations. The circuit is shown in Fig. 1. All transistors used in the circuit are designed for high-voltage operation, which is generally between 8V and 29V. The switch is turned on by pre-charging $V_{out}$ to a lower supply voltage ($V_{in1}$, which is typically 3.6V) through the two low- $V_t$, depletion-mode, n-channel (LNMOS) devices and applying 0V to the gate of the p-channel (PMOS) device. With this condition, the intermediate node ($V_{intermediate}$) is charged through the very low-$V_t$, depletion-mode, n-channel device (DNMOS) so that the $V_{gs}$ of the PMOS is larger than the magnitude of the PMOS_$V_t$. Therefore, $V_{intermediate}$ is passed to $V_{out}$, which increases the gate voltage of the DNMOS. This feedback loop eventually charges $V_{out}$ to the full level of $V_{high}$. The LNMOS devices are gated with $V_{step}$ (typically 7V) and $V_{in1}$ to prevent the increased $V_{out}$ voltage level from discharging to the $V_{in1}$ level. The simulated operation of the circuit is shown in Fig. 2. The waveforms display the voltage levels of $V_{in1}$, $V_{in2}$, $V_{intermediate}$, and $V_{out}$ while the switch is initiated to charge $V_{out}$ to the level of $V_{high}$.

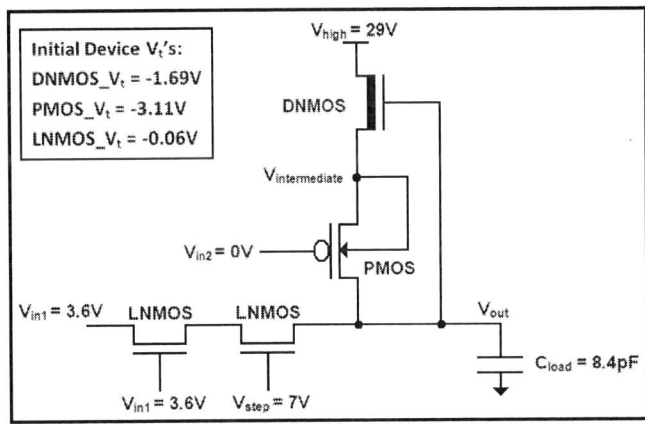

Figure 1. HV Switch with typical voltage levels designated. The purpose of the circuit is to pass the full $V_{high}$ voltage to $V_{out}$. $C_{load}$ represents a typical capacitive load connected to $V_{out}$.

978-1-4244-6572-9/10 $26.00 © 2010 IEEE

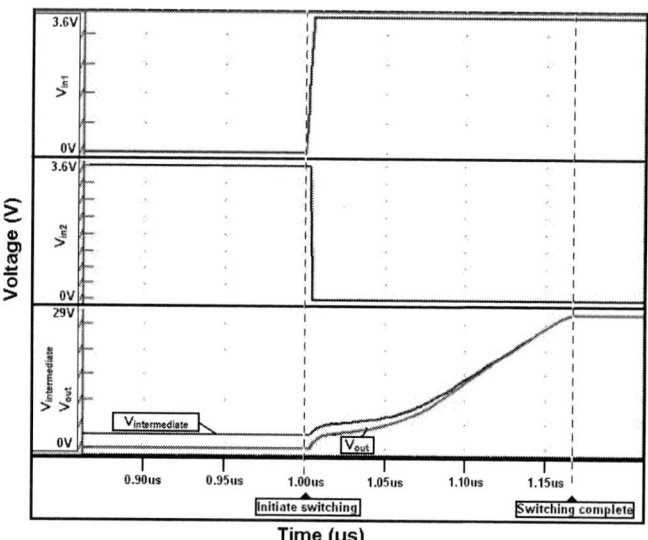

Figure 2. HSPICE simulation of HV Switch operation.

It is shown that once the switch has been turned on, the PMOS device has a bias across the gate oxide equal to the magnitude of $V_{high}$, which results in device degradation that causes an increase in $V_t$. The increased $V_t$ and concurrent decrease in drive current has the effect of slowing down the rise time of the $V_{out}$ node (tRISE) after the switching is initiated, as shown in Fig. 3. If the $V_t$ gets too large, then the switch does not turn on at all, due to $V_{gs}$ of the PMOS not being high enough. The time needed to fully charge Vout to $V_{high}$ is directly related to the magnitude of the $V_t$ shift. Therefore, by observing tRISE over operating lifetime, it is possible to compare the degradation of the circuit on the live die to the degradation of a discrete test structure. With this correlation, we are able to predict the lifetime of the circuit using test structure measurements.

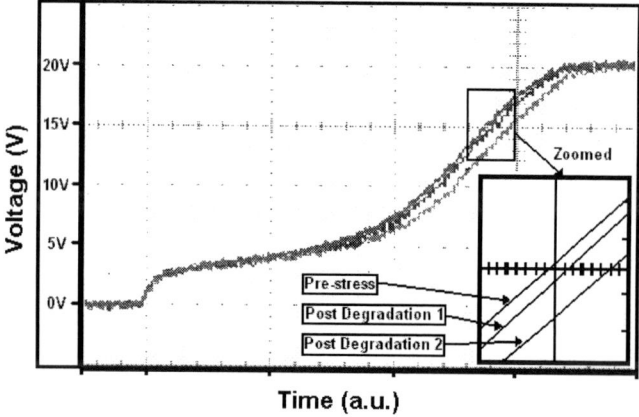

Figure 3. Oscilloscope screenshot of degradation affect on tRISE, which is increased tRISE with increased stress time. The Pre-stress curve was measured before degradation. The Post Degradation 1 curve was measured after 100 seconds of stress. The Post Degradation 2 curve was measured after 1000 seconds of stress.

## II. EXPERIMENT AND RESULTS

The discrete test structure used for degradation testing was a high-voltage surface-channel PMOS transistor. The gate oxide thickness was electrically measured to be 350A. In order to emulate use conditions in the circuit, the gate was set to 0V while the drain was varied between 0V and a high voltage (27V to 30V). The frequency of the stress waveform was set to 20 kHz, with a 30% duty cycle. The 70% portion of the waveform that has no stress bias allows relaxation of the PMOS_ $V_t$, which is the de-trapping of the charge in the oxide. The temperature during stress was 90°C. The $V_t$ of the PMOS was measured at various points in time with an automated parametric measurement station. The $V_t$ shift ($DV_t$), defined as the difference between the pre-stress $V_t$ and the $V_t$ after a specific degradation stress time, was plotted against stress time for each stress voltage level, shown in Fig. 4.

The live die testing was completed on the same wafer used to measure the test structure. Using a customized NAND flash testing platform, a macro was created that turned the switch circuit on and then back off at the same frequency and duty-cycle as the test structure test at the same conditions. The output of the switch was monitored, using a high-voltage microprobe and a 500MHz oscilloscope, to measure tRISE during switching. Measurements were performed at multiple stress times to find the relationship between tRISE and stress time. Also, $V_{high}$ was set to 20V for the tRISE measurements.

In order to correlate circuit performance to test structure measurements, SPICE simulation was used to determine the relationship between tRISE of the output of the switch to the $DV_t$ of the PMOS, using degraded device models. The relationship is dependent upon the initial $V_t$ of each device type used in the circuit, the value of the capacitive load, and the bias conditions. Therefore, the $V_t$'s of each device type were measured on the same wafer, using an automated parametric measurement station. The $V_t$ values were, then, used to model the devices in the simulation. An initial simulation was run to compare tRISE to the live die pre-stress measurement. The simulated tRISE was measured to be 158ns

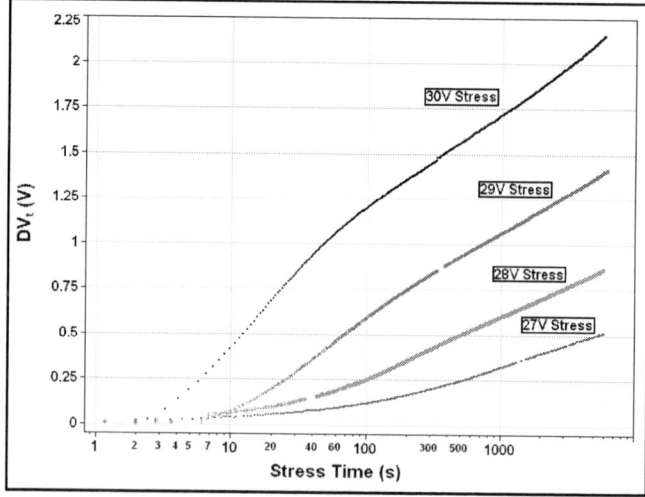

Figure 4. PMOS Test Structure $DV_t$ vs stress time by stress voltage level.

and the live die was measured to be 160ns. Subsequent simulations were run with the PMOS_$V_t$ swept (increasing in magnitude) to measure the impact of the device degradation on tRISE. From the simulation results, an empirical fit was used to determine the equation relating $DV_t$ to tRISE:

$$DV_t = Ln(\alpha + \beta * Ln(tRISE)) \tag{1}$$

where Ln is the natural log, and constants $\alpha$ and $\beta$, equal to 78.92 and 4.97, were determined by iterative fitting, using statistical software. The values of the constants are dependent upon the bias conditions, the capacitive load, and the $V_t$'s of the three device types.

Using eq. (1), the relationship between $DV_t$ and stress time was determined for the live die. The $DV_t$ vs stress time was plotted and compared to the test structure results, shown in Fig. 5. It is observed that the difference between live die and test structure is less than 3.5%, as shown in Table 1.

## III. CONCLUSIONS

The strong correlation between the test structure degradation rate and circuit operation over time confirms the accuracy of the developed methodology. This method enables the prediction of circuit lifetime using test structure measurements and SPICE simulation with degraded device models. The described practice can be applied to any circuit where a timing metric is related to an understood degradation mechanism. The results of this correlation method could also be used to predict circuit performance with non-degraded test structure measurements if device parameters are shifted due to process variation.

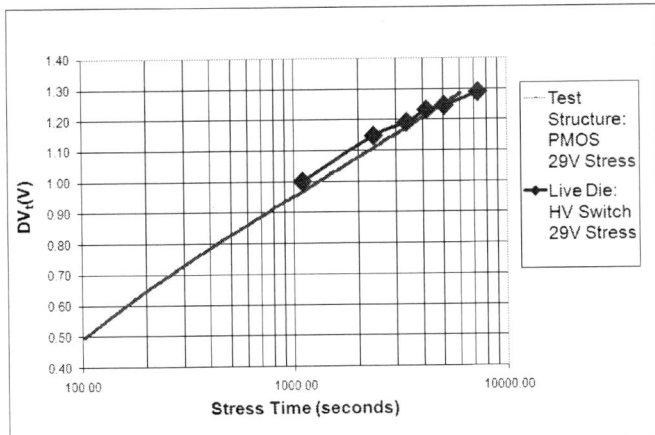

Figure 5. PMOS $DV_t$ vs Stress Time for test structure and live die.

TABLE I.  LIVE DIE AND TEST STRUCTURE PMOS $DV_T$ VALUES ARE COMPARED. LIVE DIE $DV_T$ IS WITHIN 3.5% OF TEST STRUCTURE VALUES.

| Stress Times (s) | Live Die tRISE (ns) | From Simulation Calculated $DV_t$ (V) | Test Structure Measured $DV_t$ (V) | % Difference |
|---|---|---|---|---|
| 1098 | 220 | 1.00 | 0.97 | 2.97% |
| 2364 | 240 | 1.15 | 1.11 | 3.26% |
| 3379 | 247 | 1.19 | 1.17 | 1.68% |
| 4181 | 253 | 1.23 | 1.22 | 0.81% |
| 5064 | 256 | 1.24 | 1.25 | -0.45% |
| 7296 | 264 | 1.29 | 1.32 | -2.33% |

## IV. REFERENCES

[1] G. Lawton, "Improved flash memory grows in popularity," in IEEE Computer Society. Vol. 39. Issue 1. Jan. 2006. pp. 16-18.

[2] D. Nobunaga, et al., "A 50nm 8Gb NAND flash memory with 100MB/s program throughput and 200MB/s DDR interface," in IEEE ISSCC Digest of Technical Papers. Feb. 2008. pp. 426-625.

[3] T. Cho, et al., "A 3.3 V 1 Gb multi-level NAND flash memory with nonuniform threshold voltage distribution," in IEEE ISSCC Digest of Technical Papers. Feb. 2001. pp. 28–29.

[4] T. S. Jung, et al., "A 3.3V 128Mb multi-level NAND flash memory for mass storage applications," in IEEE ISSCC Digest of Technical Papers, Feb. 1996. pp. 32-33.

[5] M. Bauer, et al., "A multilevel-cell 32Mb flash memory," in IEEE ISSCC Digest of Technical Papers. Feb. 1995. pp. 132-133.

[6] P. Samanta, C.K. Sarkar, "New positive charge trapping dynamics in SiO2 gate oxide, based on bulk impact ionization processes under Fowler-Nordheim stress," in Microelectronics Reliability. Vol. 38. 1998. pp. 1969-1973.

[7] P. Samanta, C.K. Sarkar, "Gate oxide degradation due to trapping of positive charges during Fowler-Nordheim stress at low electron fluence: a rigorous model," in International Conference of Microelectronics. Vol. 2. Sept. 1997. pp. 14-17.

[8] K. Kazushige, et al., "A 120mm2 16GB 4-MLC NAND flash memory with 43nm CMOS technology," in IEEE ISSCC Digest of Technical Papers. Feb. 2008. pp. 430-431, 625.

# A Comprehensive Study on Nanomechanical Properties of Various SiO$_2$-based Dielectric Films

Guohua Wei, Sony Varghese, Kevin Beaman, Irina Vasilyeva, Tom Mendiola,
Andrew Carswell, David Fillmore and Shifeng Lu

Micron Technology, Inc., 8000 S. Federal Way, Boise, Idaho, USA

*Abstract*—This paper studies the nanomechanical properties, including hardness and Young's modulus (both in a dry condition and in deionized water), fracture toughness, cohesive strength and scratch resistance of eight commonly used SiO$_2$-based dielectric films, Boron Phosphosilicate Glass (BPSG), BPSG with Rapid Thermal Processing (RTP), Phosphosilicate Glass (PSG), Spin-On Dielectric (SOD), Plasma Enhanced Chemical Vapor Deposition (PECVD) Tetraethyl Orthosilicate (TEOS), high aspect ratio oxide (O3-TEOS), High Density Plasma Oxide (HDP), and Silane Oxide. Significant differences were found among these films. The effects of the nanomechanical properties on dielectric film reliability and CMP process are discussed.

*Keywords-silicon dioxide; dielectric films; nanoindentation; nanoscratch; fracture toughness; cohesive strength; Chemical Mechanical Planarization (CMP)*

## I.    INTRODUCTION

The mechanical strength of SiO$_2$-based dielectric films used in semiconductor production plays an important role in dielectric film reliability [1-2]. The mechanical failures such as cohesive fracture of the dielectric films can occur when mechanical or thermal stresses are introduced during process integration, in which a wide range of SiO$_2$-based dielectric films are used, depending on the requirements for the specific process. A thorough study of the nanomechanical properties of these films is desired for process development and process integration.

The nanomechanical properties of SiO$_2$-based dielectric films are also important for Chemical Mechanical Planarization (CMP) process development [3-4]. Many of the dielectric films will go through the CMP process. In order to develop the appropriate slurries for SiO$_2$-based dielectric films, it is necessary to study the relationship between CMP removal rate and nanomechanical properties of dielectric films. In addition, a systematic study of nanomechanical properties, including scratch resistance, of various dielectric films should also help to better understand and deal with the CMP defects such as nanoscratches (~ several microns long, 5-20 nm deep).

This paper presents a comprehensive study on nanomechanical properties of eight SiO$_2$-based dielectric films, Boron Phosphosilicate Glass (BPSG), BPSG with Rapid Thermal Processing (RTP) reflow, Phosphosilicate Glass (PSG), Spin-On Dielectric (SOD), Plasma Enhanced Chemical Vapor Deposition (PECVD) Tetraethyl Orthosilicate (TEOS), high aspect ratio process oxide (O3-TEOS), High Density Plasma Oxide (HDP), and Silane Oxide, which are commonly used in semiconductor manufacturing. The hardness, Young's modulus, fracture toughness, cohesive strength and scratch

resistance of these films were measured, and their effects on dielectric reliability and CMP process are discussed. The hardness and Young's modulus were also measured in wet medium (deionized (DI) water) to better simulate the CMP condition to study the impact of oxide hydrolysis on nanomechanical properties and CMP.

## II.    EXPERIMENTAL DETAILS

The eight dielectric films were deposited on 300 mm Si (100) wafers. Two sets of wafers were prepared; one set for nanomechanical property studies and the other set for CMP removal rate measurement. Table I lists the thickness and main applications of the films studied in this paper.

The hardness, Young's modulus, fracture toughness and scratch resistance of the dielectric films were all measured by the Nano Indenter® XP system. The hardness and Young's modulus were measured at both the dry condition (no wet medium involved) and in DI water using the Continuous Stiffness Measurement (CSM) technique with a Berkovich tip (tip radius < 20 nm, total included angle = 142.5°) [5]. To do the test in DI water, a drop of DI water was put on the surface of the oxide film and the hardness and modulus were measured [6]. The fracture toughness was measured with a three-sided Cube Corner tip (tip radius < 20 nm, total included angle = 90°). A peak indentation load of 5.0 mN was applied to create the indentation-induced cracks, and the crack lengths were then measured by a JEOL 6340 Scanning Electron Microscope (SEM). The scratch resistance of the dielectric films was measured using the "constant loading nanoscratch" technique at 1.0 mN with another Berkovich tip, which was aligned to scratch the surface in a face-forward direction. The

TABLE I.    AN OVERVIEW OF SAMPLE INFROMATION

| Oxide type | Thickness (nm) | Main applications |
|---|---|---|
| BPSG | 800 | To place thick dielectric layers between lines or contacts |
| BPSG with reflow | 950 | To harden the BPSG to lower the etch rate and improve gap-fill |
| PSG | 1450 | To form the DRAM capacitor level |
| SOD | 600 | To fill high aspect ratio structures with no voids or seams |
| O3-TEOS | 720 | To fill high aspect ratio structures particularly when SOD integration is difficult |
| HDP | 700 | To fill high aspect ratio structures. Gap-fill ability not as good as SOD and HARP |
| PECVD TEOS | 700 | To be used when a low temperature oxide is required |
| Silane oxide | 700 | To be used in BEOL applications for very thick, low temperature requirement |

details of the nanoscratch technique can be found elsewhere [7].

In order to do the CMP removal rate measurement, all the wafers were processed on an Applied Materials Reflexion LK 300 mm polisher. They were run as a two platen process on Dow IC1010 pads followed by a water buff on a soft pad on platen 3. Commercially available colloidal silica slurry was used. The process conditions were the following – 3 psi average down force on IC1010 pads, platen speed 47 rpm, carrier speed 63rpm and polish time 35s on each of the 2 platens. Wafers were then cleaned using dilute HF in the polisher cleaners. The wafer thicknesses were measured on a NOVAScan 3090 optical metrology tool. Removal rates were calculated as the oxide removed during the total 70s polish (oxide removed by the dilute HF cleans were accounted for in the removal rate calculations).

## III. RESULTS AND DISCUSSION

### A. Hardness and Young's modulus

Fig. 1 shows the hardness and Young's modulus of eight SiO$_2$-based dielectric films measured in both the dry condition and in DI water. Although they are all SiO$_2$-based films, significant hardness and modulus difference is seen. The HDP, TEOS and SOD oxides have the highest hardness and modulus, while the PSG shows the lowest hardness and modulus. The hardness of HDP, SOD and TEOS is almost twice that of PSG. It is interesting to see that the doped silicon oxides, PSG and BPSG, show relatively lower hardness and modulus than the un-doped ones, except O3_TEOS, which shows lower hardness and modulus than BPSG and BPSG with reflow. The BPSG reflow shows higher hardness and modulus than BPSG. This is expected, since the main purpose of the reflow is to harden the BPSG films. The data variation of silane oxide is larger than the other oxide films, because the

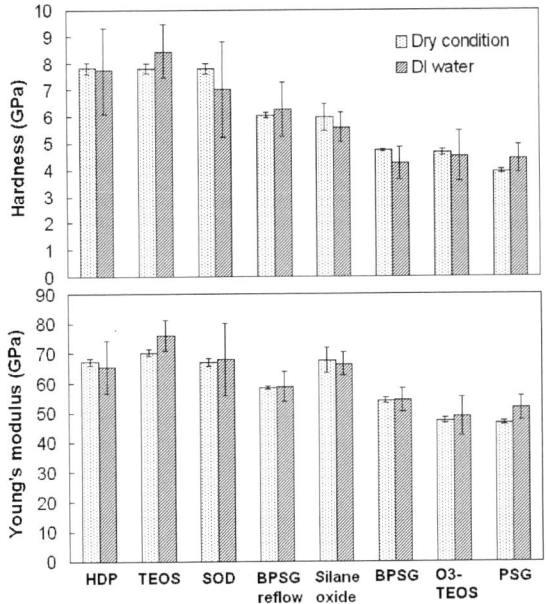

Figure 1. Bar chart of hardness and Young's modulus of eight oxide films measured in dry conditions and DI water.

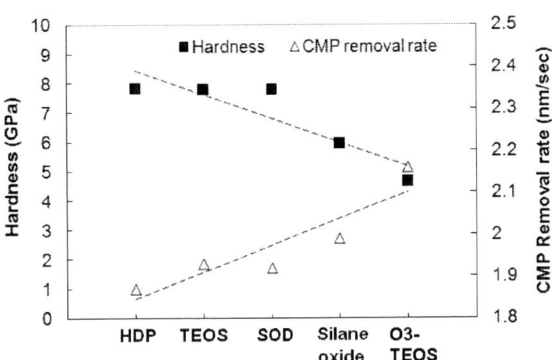

Figure 2. The relationship between hardness and CMP removal rate for five un-doped oxide films.

surface of silane oxide was much rougher than the other films, which would affect the nanoindentation measurement repeatability.

In DI water, no significant change was found for the hardness and modulus compared to dry conditions. However, the data variation increased significantly (higher standard deviation error bar) for all the oxide films except the silane oxide, which showed similarly high data variation in both dry condition and in DI water. The reason for the data variation change in DI water is not clear. It is speculated that the surface tension of the DI water might affect the indentation load to some extent during the measurement, causing the data to scatter.

Fig. 2 shows the relationship between the hardness and CMP removal rate for five un-doped oxide films. A good correlation between hardness and CMP removal rate is observed. That is, the lower the hardness, the higher the CMP removal rate. The CMP removal rates of three doped oxide films, BPSG with reflow (7.78 nm/sec), BPSG (7.91 nm/sec) and PSG (6.77 nm/sec) are significantly higher than the un-doped oxide films, and the data is not shown in Fig. 2, since the magnitude of the increase of their CMP removal rates is not attributed to hardness decrease. It is well known that doping an oxide films with boron or phosphorus increases the CMP removal rate significantly [8]. A follow up paper will discuss results from CMP in more detail.

### B. Fracture toughness and cohesive strength

Fracture toughness, $K_c$, is the resistance of a material to failure from fracture starting from a preexisting crack, and it was determined by the following equation [9]:

$$K_c = \beta \left( \frac{E}{H} \right)^{1/2} \frac{P}{c^{3/2}} \quad (1)$$

where $P$ is the maximum indentation load (5.0 mN), $E$ is the elastic modulus measured in Fig. 1, $H$ is the hardness shown in Fig. 1, $c$ is the crack length measured by SEM, and $\beta$ is the empirical constant which depends on the geometry of the indenter tip. The value of $\beta$ is 0.032 for the cube corner tip used in this work.

Fig. 3 shows the SEM images of the representative indent made on each oxide. The indentation-induced cracks that extended beyond the indent corner can be clearly seen on six oxide films except the silane oxide and BPSG. Long indentation-induced cracks may correspond to low fracture toughness according to (1).

Fig. 4 (a) shows the bar chart of fracture toughness of six oxide films. Although the fracture toughness of silane oxide and BPSG could not be calculated and is not shown, they are believed to have higher fracture toughness than the other films because no extended cracks were induced for these films under the same load indentation test. The HARP shows the lowest fracture toughness. It is interesting to note that although the BPSG reflow has higher hardness and modulus than BPSG, its fracture toughness is lower than BPSG. This is because an increase of hardness also increases the brittleness of the material, which may lead to lower fracture toughness. This also explains why HDP, TEOS and SOD which show the highest hardness and modulus, do not have the highest fracture toughness. Therefore, fracture toughness is an important material property in addition to hardness and Young's modulus that deserves more attention, especially when dealing with the cracking issue of the oxide films.

It should be noted that the calculation of fracture toughness could be affected by the residual stresses of the oxide films. The tensile stress may facilitate the propagation of the crack tip, while the compressive stress may prohibit the crack tip propagation, which could impact the value of the toughness deduced from the crack length [9, 10]. However, it is known from the manufacturing that BPSG, PSG and O3-TEOS usually have similar tensile stresses, but the fracture toughness of these three films is very different. In addition, the HDP usually has higher compressive stress than silane oxide, but the HDP has lower fracture toughness than silane oxide. So it appears that the residual stress was not a dominating factor for fracture toughness measurement in this study.

The mechanical robustness of films can be characterized by cohesive strength (fracture energy), which represents the energy required to break all the chemical bonds in the path of a propagating crack, and it was determined by the following equation [1, 6]:

$$J_{cohesive} = \frac{K_c^2}{E} \qquad (2)$$

where $K_c$ is the fracture toughness measured in Fig. 4 (a) and $E$ is the Young's modulus measured in Fig. 1.

Fig. 4 (b) shows the cohesive strength of six oxide films except silane oxide and BPSG. Since the fracture toughness of silane oxide and BPSG was not available, their cohesive strength could not be calculated. It can be seen that the BPSG with reflow has the highest cohesive strength, while the HARP and SOD show the lowest cohesive strength. It is worth noting that both the O3-TEOS and SOD are used to fill high aspect ratio structures. Although they have superior gap-fill ability compared to HDP oxide, their mechanical robustness is not as good as HDP. This may be a concern for the device reliability under high mechanical or thermal stresses for O3-TEOS and SOD. When the dielectric film stress is high enough that the strain release energy exceeds the cohesive strength, the spontaneous cracking of the films will occur [6].

Figure 3. SEM images of the representative indent made on each oxide sample showing the indentation-induced cracks.

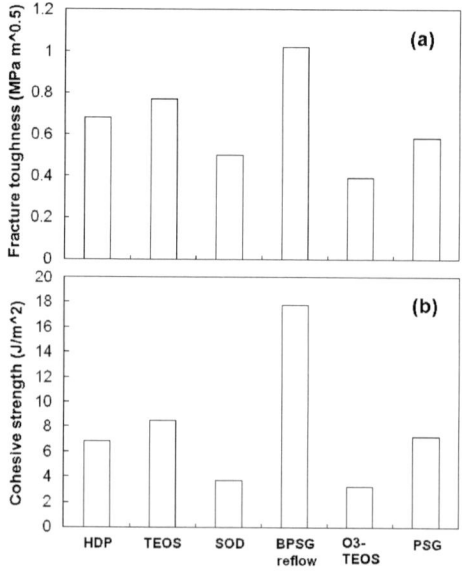

Figure 4. Bar chart of fracture toughness (a) and cohesive strength (b) of six oxide films.

*C. Scratch resistance*

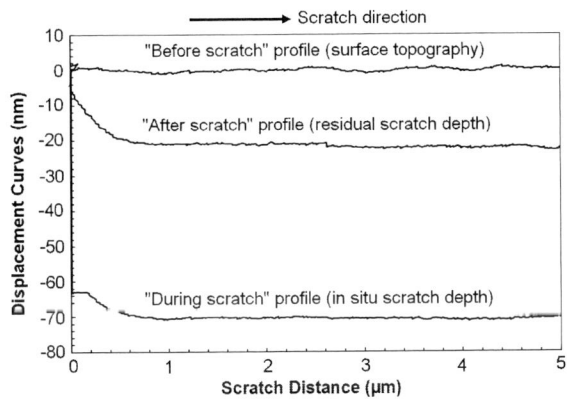

Figure 5. Example of scratch profiles obtained during a 1.0 mN constant loading nanoscratch test. The data was from the HARP oxide.

Fig. 5 shows the typical scratch depth profiles obtained during the nanoscratch test. Three profiles can be seen. The "before scratch" profile shows the topography of the sample. The "during scratch" profile was obtained during the 1.0 mN scratch, and it shows the "*in situ*" scratch depth caused by the diamond tip. The "after scratch" profile was obtained after the scratch test, and it showed the "residual depth" of the sample, which reflects the extent of permanent damage and plowing of the diamond tip into the sample. The difference between the *in situ* and residual depth profiles is attributed to elastic recovery after removal of the normal load [7].

Fig. 6 shows the bar chart comparing the *in situ* and residual scratch depths of eight oxide films obtained during the constant loading nanoscratch test. The residual scratch depths are in the range of 5-20 nm, which correspond to the depths of the scratch defects found on oxide films during CMP according to Atomic Force Microscope (AFM) data, which is not shown in the paper. The HDP, TEOS and SOD oxides show the lowest in situ scratch depths (< 50 nm), while the PSG, HARP and BPSG show the highest in situ scratch depth (> 60 nm), which exhibits a good correlation to the hardness data. By comparing the residual depths, the HDP, TEOS and BPSG with reflow seem to show the lowest values (< 10 nm), while the PSG, O3-TEOS and silane oxide are the three oxides that show the highest residual depth (~ 20 nm). Under the same scratch load, a lower residual scratch depth usually corresponds to better scratch resistance.

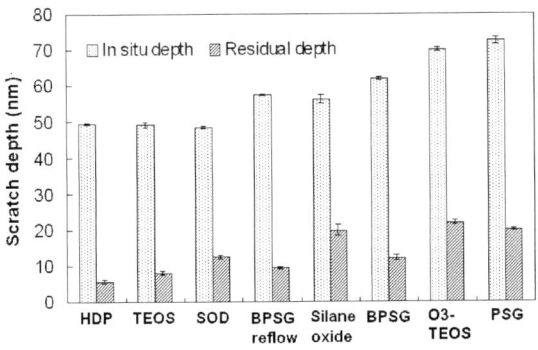

Figure 6. Bar chart of *in situ* and residual scratch depths of eight oxide films.

## IV. CONCLUSIONS

In this study, it is found that the hardness of HDP, TEOS and SOD is almost twice that of PSG. The hardness and Young's modulus of un-doped oxides (HDP, TEOS, SOD, silane oxide) are generally higher than the doped oxides (BPSG, PSG), except O3-TEOS which shows low hardness and Young's modulus. The effect of DI water on hardness and modulus of the oxide films studied in the paper was found to be insignificant. A good correlation between hardness and CMP removal rate was found for un-doped oxides. The doped-oxides (BPSG, BPSG with reflow, PSG) show significantly higher CMP removal rates than the un-doped oxides. This can be attributed to the effects of doping, not the hardness decrease. The silane oxide and BPSG show the highest fracture toughness. The SOD and O3-TEOS show the lowest fracture toughness and cohesive strength. The HDP, TEOS and BPSG with reflow show the best scratch resistance, and the PSG, O3-TEOS and silane oxide show the least scratch resistance.

## ACKNOWLEDGMENT

The authors would like to thank Radha Padmanabhan, Suresh Ramakrishnan, Deanna King, Zabit Halla, Matt Caldwell, Rachel Ruz, Joe Wiggins, Scott York, Jim Jozwiak, Ariela Gruszka, Mike Benson and Arthur McGinnis for their support and useful discussions during this study.

## REFERENCES

[1] S. W. King, J. A. Gradner, "Intrinsic stress fracture energy measurements for PECVD thin films in the $SiO_xC_yN_z$:H system", Microelectron. Reliab., vol. 49, pp. 721-726, 2009.

[2] G. Xu, J. He, E. Andideh, J. Bielefeld and T. Scherban, "Cohesive strength characterization of brittle low k films", Proc. of the IEEE 2002 IITC, pp. 57-59, 2002.

[3] P. B. Zantyea, A. Kumara and A.K. Sikderb, "Chemical mechanical planarization for microelectronics applications", Mater. Sci. Eng. R. vol. 45, pp. 89-220, 2004.

[4] P. Leduc, T. Farjot, M. Savoye, A.Demas, S. Maitrejean, G. Passemard, "Dependence of CMP-induced delamination on number of low-k dielectric films stacked", Microelectron. Eng., vol. 83, pp. 2072-2076, 2006.

[5] W. C. Oliver, G. M. Pharr, "An improved technique for determining hardness and elastic modulus using load and displacement sensing indentation experiments", J. Mater. Res., vol. 7, pp. 1564-1583, 1992.

[6] E. E. Simonyi, M. Lane, E. Liniger and A. Grill, "Comparison of the fracture behavior of brittle ILD films used in the BEOL in dry and wet environment using nanoindentation", Mater. Res. Soc. Symp. Proc., vol. 914, pp. 389-394, 2006.

[7] G. Wei, M. L. Weaver and J. A. Barnard, "Nanotribological studies of chromium thin films", Tribology Letters, vol. 13, pp. 255-261, 2002.

[8] B. A. Bonner, B. Fishkin, J. David, C. Garretson and T. H. Osterheld, "Removal rate, uniformity and defectivity studies of chemical mechanical polishing of BPSG films", Mater. Res. Soc. Symp. Proc., vol. 613, E8.6.1-E8.6.6, 2000.

[9] G. M. Pharr, D. S. Harding and W. C. Oliver, "Measurement of fracture toughness in thin films and small volums using nanoindentation methods", in Mechanical Properties and Deformation Behavior of Materials Having Ultra-Fine Microstructures, M. Nastasi, D. M. Parkin and H. Gleiter, Eds. Kluwer Academic, pp. 449-461, 1993.

[10] M. Ciccotti, "Stress-corrosion mechanisms in silicate glasses", J. Phys. D: Appl. Phys., vol. 42, pp. 214006, 2009.

978-1-4244-6572-9/10 $26.00 © 2010 IEEE

# A Novel Depletion Mode High Voltage Isolation Device

Vladimir Mikhalev and Michael Smith

Micron Technology Inc., Department of Process R&D

8000 S. Federal Way, Boise, USA

masmith@micron.com

*Abstract*—**A novel depletion mode high voltage isolation device is presented. It consists of a narrow n- resistor covered with a grounded metal field plate. This device will pass low voltages, but will block high voltages. It has potential application to isolate a NAND memory array from periphery low voltage circuitry, and has the benefit that it can be made more compact than a standard MOSFET device and can be integrated into the process without adding or changing process steps.**

*Keywords: NAND flash, depletion mode, high voltage isolation*

## I. INTRODUCTION

The periphery MOSFETs that supply high operation voltages (20-35V) to the NAND memory array are shrink limited by their voltage handling requirements and have not significantly shrunk with the scaling of the memory array. High voltage (HV) devices which are pitched with the array provide an opportunity to impact die size if their dimensions and/or spacing can be shrunk without compromising HV functionality.

The device which isolates the memory array from the dynamic data cache (DDC) is one such pitched device. During array erase operation the bitlines will float up to the erase voltage (~20V), requiring that the sensitive low voltage (LV) circuitry in the DDC be isolated from the array. But during other array operations, such as read the DDC needs to be able to pass low voltages to the array. The standard solution is to place a HV MOSFET, with robust contact spacing design rules to allow for high breakdown voltage, between the array and DDC. During erase this FET is turned off ($V_g$=0V) to block the array side HV, and during LV operations the device is turned on ($V_g$>0V) to allow LV to pass through to the array.

The proposed novel device may serve as a replacement for the standard HV isolation MOSFET. It has the advantages of being more compact than a standard MOSFET, handles the required high voltages, and can be integrated into the existing process flow without the addition or alteration of steps. It can be described as a narrow depletion-mode metal gated FET, although in the proposed application the gate would be grounded and serve as a field plate so that the device would function as a diffusion resistor with a process and layout definable depletion voltage ($V_{depl}$). The low n- doping, the

presence of the metal field plate, and the field fringing along the STI isolation edge combine to allow for a relatively low $V_{depl}$, which serves as the critical voltage above which the applied voltages will not be passed to the output node.

## II. DEVICE DESCRIPTION

The device is illustrated in Fig.1. It consists of a narrow lightly doped n- channel between two contact nodes, which are wider than the channel as necessary to allow for high contact to active area edge breakdown. In general, the device may be asymmetrical to minimize size, with the LV side contact surround smaller than the HV side. The contact heads may

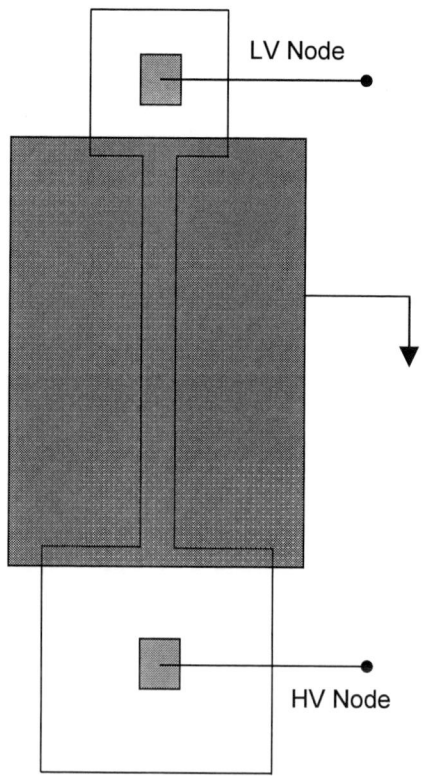

Figure 1. Top down view of device. Channel is very narrow, and contact landing pads are sized to allow for high breakdown voltage. The grounded field plate covers the channel.

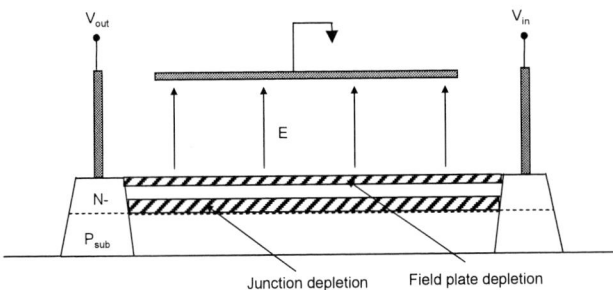

Figure 2. Depletion of n- channel from vertical p-n junction and grounded field plate. Full depletion occurs when the two depletion regions merge.

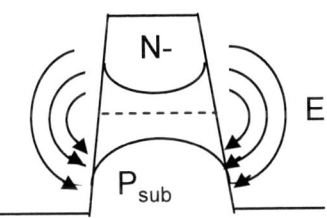

Figure 3. Field fringing of the vertical p-n junction becomes a significant factor when the active area width is small. By using a very narrow channel (~100nm), the depletion of the n- channel is greatly enhanced, and full depletion occurs at lower voltages.

have higher doping as necessary to minimize resistance while still allowing for high breakdown voltage. A metal field plate covers the channel area, and in the proposed application will be grounded, although in theory it could be forced to another voltage to fine tune $V_{depl}$. The field plate is fabricated using the first metal layer, which is about 5000Å above the silicon surface.

The light n- doping, narrow channel, and presence of the field plate result in a relatively low $V_{depl}$, which can be tuned through layout variables (channel width, field plate design), process variables (n- doping level), and possibly circuit variables (field plate set to non-zero voltage). $V_{depl}$ is the key parameter of this device and must be set within a window defined by the LV pass and HV block requirements as discussed below. The above variables can be tuned to set $V_{depl}$ as required, and the presence of multiple variables provides flexibility so that the device can be integrated into an existing process flow without cost-adding changes, such as a dedicated implant.

In operation, the device passes applied voltages ($V_{in}$) below $V_{depl}$, but significantly blocks higher voltages. For applied voltages below and up to $V_{depl}$, the channel does not deplete and the output node is conductively coupled to the input node, and if allowed to will float up to the input voltage. For applied voltages above $V_{depl}$, the output voltage will increase above

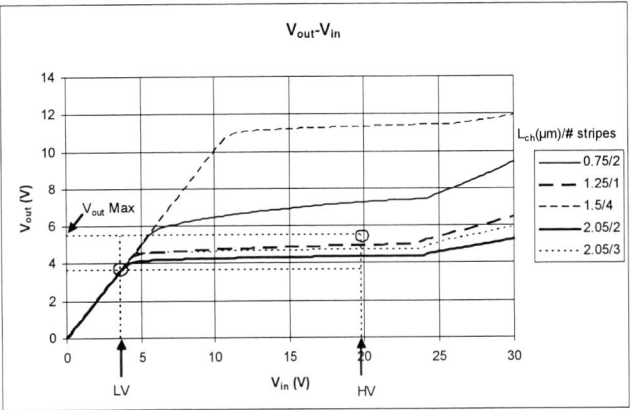

Figure 4. Voltage pass characteristics of various sized devices. $V_b=V_g=0V$, $I_{out}=-10nA$. $V_{out}$ follows $V_{in}$ until the channel fully depletes, then $V_{out}$ increases slowly, depending on the electrostatic coupling. Device with 4 stripes (labeled 1.5/4) has high $V_{depl}$ due to proximity of neighboring stripes causing frustration of field fringing. 1.25/1 is overall the best device.

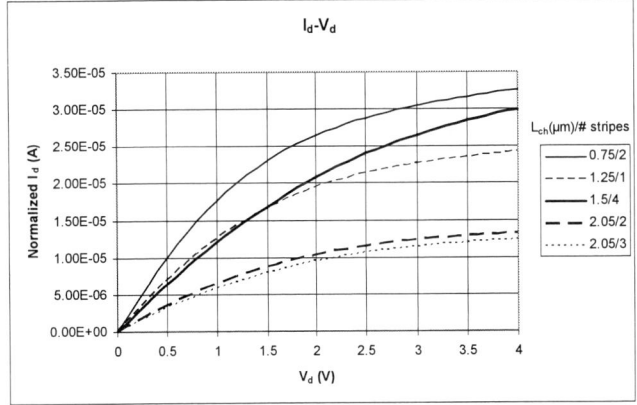

Figure 5. $I_d$-$V_d$ of devices with source grounded. $I_d$ normalized to number of stripes, which is indicated in key.

$V_{depl}$ by only a fraction of $V_{in}$-$V_{depl}$ due to electrostatic coupling along the depleted channel. Ideally, the coupling factor would be zero so that the output would never exceed $V_{depl}$, but in reality the coupling factor is inversely proportional to the channel length ($L_{ch}$), thus making $L_{ch}$ a critical design parameter.

This device utilizes three known depletion mechanisms associated with RESURF devices. As shown in Fig. 2, the first mechanism is depletion of the vertical n-p junction of the channel [1]. Secondly, the presence of the grounded field plate depletes the channel at the surface, thus lowering $V_{depl}$. [2]. $V_{depl}$ is further lowered by using narrow active area stripes to create the dielectric RESURF effect, which enhances depletion via field fringing at the STI edges as illustrated in Fig. 3 [3]. For the NAND array isolation application, it was found that only one stripe is necessary to provide sufficient channel conductivity, and allows for the smallest device size.

III. MEASUREMENTS AND DATA

Test structures with various device sizes, including channel length and number of stripes, were fabricated and measured for I-V characteristics. The parametrics of special interest were $V_{depl}$, $V_{out}/V_{in}$ slope for $V_{in} > V_{depl}$ (i.e. coupling factor), and resistance for $V_{in} < V_{depl}$. The results are shown in Figs. 4, 5, and 7 (all data presented here taken from devices with stripe width = 100nm). A simulation of the effect of $L_{ch}$ on $V_{out}$-$V_{in}$ relation is shown in Fig. 6.

Fig. 4 shows the $V_{out}$-$V_{in}$ curves. $V_{depl}$ is evident from the knee in the curve, below which the $V_{out}$-$V_{in}$ slope is 1, and above which the slope is a function of $L_{ch}$. The kink that appears near 25V is related to breakdown of the HV junction. The reference lines indicate one particular set of requirements: $V_{in} = 3.6V \rightarrow V_{out} = 3.6V$, and $V_{in} = 20V \rightarrow V_{out} < 5.5V$. For the first requirement to be met, $V_{depl}$ must be larger than 3.6V. For the second requirement to be met, $V_{depl}$ must be less than 5.5V, and the $V_{out}/V_{in}$ slope must be shallow enough so that Vout does not increase above 5.5V for Vin = 20V. Note that for zero slope, the window for $V_{depl}$ is 3.6V to 5.5V. For non-zero slopes, this window shrinks with increasing slope.

Fig. 7 shows the slope-v-$L_{ch}$ curves taken from the measurements and simulation. Fitting a power curve to these data results in an exponent of approximately -2 for the measurements and -2.5 for the simulation. Evidently this uncalibrated 2D simulation overestimates the slope. The lower limit of $L_{ch}$ is set by the $V_{depl}$ window requirement. To estimate an absolute limit for $L_{ch}$, one can solve the fitted power curve equation for $V_{out}/V_{in}$ slope = 0.12 (using delta $V_{out}$ = 5.5V-3.6V, delta $V_{in}$ = 20V-3.6V), giving $L_{ch} = 0.62\mu m$. For comparison, this value is smaller than a typical contact-to-gate design rule or gate length, thus the total length of the device can be made significantly shorter than a standard equivalent MOSFET.

## IV.  CONCLUSIONS

A novel device is presented that may serve as a replacement of the standard NAND array high voltage isolation MOSFET. The new device can be made more compact, meets the high voltage handling requirements, and can be integrated into existing process flows without cost-adding step changes. The critical parameters of the device are the depletion voltage, which is a function of several layout and process parameters, and the $V_{in}$-$V_{out}$ electrostatic coupling factor, which is approximately proportional to the inverse square gate length. To meet typical NAND voltage requirements, an absolute limit of gate length is estimated to be about 0.62µm.

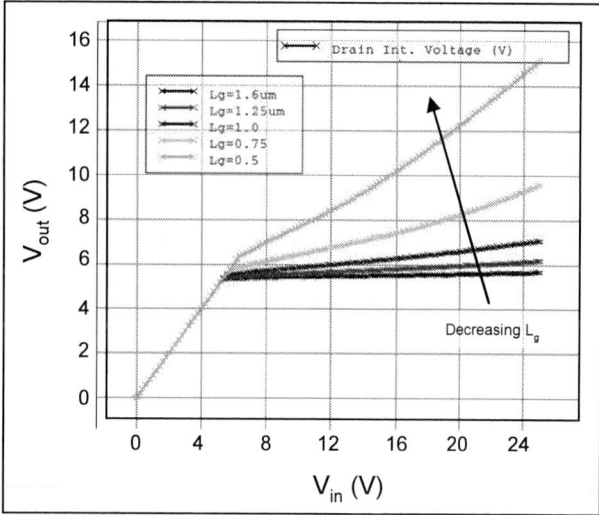

Figure 6.  Simulation of voltage pass characteristics for various channel lengths. Note that $V_{out}/V_{in}$ slope in depletion region increases with shorter gate length.

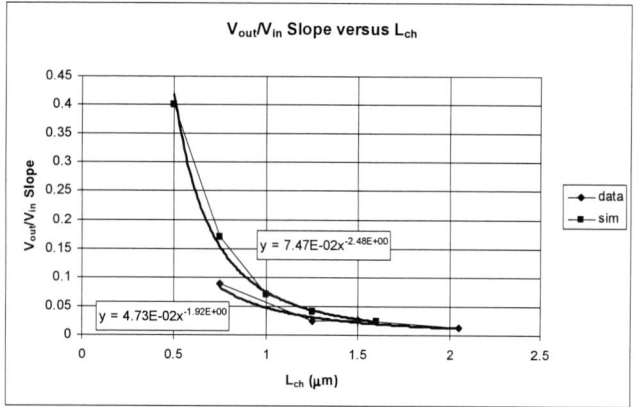

Figure 7.  $V_{out}/V_{in}$ slope in depletion region plotted against channel length for measured data and simulation. Simulation shows larger increase in slope with decreasing channel length than does data. Slope is approximately proportional to inverse $L_{ch}$ squared.

### ACKNOWLEDGMENTS

The authors would like to thank Haitao Liu for providing simulations.

### REFERENCES

[1]  A. Ludikhuize, "A review of RESURF technology," ISPSD 2000, pp. 11-18, May 2000

[2]  T. Okabe, I. Yoshida, S. Ochi, S. Nishida, M. Nagata, "A complementary pair of planar-power MOSFET's," IEEE Transactions on Electron Devices, vol. ED-27, no. 2, February 1980

[3]  J. Sonsky, A. Heringa, "Dielectric resurf: breakdown voltage control by STI layout in standard CMOS," Electron Devices Meeting, 2005 IEDM Technical Digest, pp. 373-376, December 2005

978-1-4244-6572-9/10 $26.00 © 2010 IEEE

# Fullband Study of Ultra-scaled Electron and Hole SiGe Nanowire FETs

Abhijeet Paul*, Saumitra Mehrotra, Mathieu Luisier and Gerhard Klimeck

School of Electrical and Computer Engineering, Network for Computational Nanotechnology, Purdue University,
West Lafayette, Indiana - 47907, USA.
Email: paul1@purdue.edu

*Abstract*— **Ultra-scaled SiGe nanowire FETs (NWFETs) are an attractive candidate in achieving faster p-type devices compared to Silicon. This work investigates the performance of SiGe nanowire FETs (NWFETs) using a Virtual Crystal Approximation (VCA) method based on an atomistic Tight-Binding (TB) model. The electronic structure calculation is self-consistently coupled to a 2D Poisson solver. The spatial charge and current distribution in these NWFETs strongly depend on the Ge% as well as on the channel orientation. We predict an improvement in both SiGe n and p FETs in terms of $I_{ON}$ and gate delay ($\tau_D$) compared to Silicon. For Ge > 80% the <110> oriented channels show better performance compared to the <100> SiGe n and p-FETs .**

**Keywords- MOSFETs; SiGe; nanowire; Tight-Binding; VCA**

## I. INTRODUCTION

The continuous down-scaling of the channel length of Si MOSFETs according to the ITRS [1] requirements have made it necessary to look for alternative device solutions. This includes newer device architectures like nanowire-FETs (NWFETs), finFETs, etc., as well as new channel materials. SiGe NWFETs provide a viable alternative due to their high channel mobility for both electrons and holes [2 - 4] as well as better electrostatic gate control over the channel which helps reduce the short channel effects (SCEs) [4]. A schematic of a SiGe NWFET is shown in Fig. 1a with oxide (SiO$_2$ as well as High-κ) grown all around. Below is also shown the variation of the conduction band (CB) edge Ec and valence band (VB) edge Ev in the cross-section of the wire.

Recent development in the process technology has enabled the fabrication of high Ge% ultra-scaled SiGe NWFETs [2 - 4]. Analysis of such highly scaled devices can be enabled only by a simulator which simultaneously handles the material, the strain, the quantum confinement variation and the electron and hole band coupling. This work presents the results of self-consistent calculations for SiGe NWFETs using a generic atomistic Tight-Binding based VCA method (TB-VCA).
The paper is organized as follows, Section II provides a brief description of the numerical procedure adopted to obtain the electronic structure and the self-consistent solution as well as the device details. Section III A discusses the CB and VB bandstructure in these NWFETs. Section III B and C focus on the factors which affect the charge and current distribution respectively, in these NWFETs. Section III D compares the transfer characteristics and the device performance of n and p SiGe NWFETs. Conclusions are outlined in Section IV.

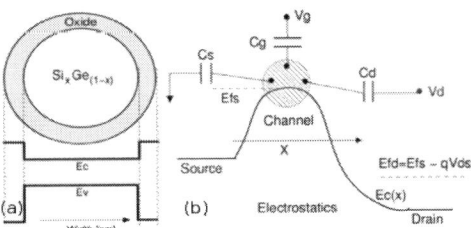

**Figure 1.** (a) Schematic of a SiGe NWFET. Core is Si$_x$Ge$_{(1-x)}$ alloy with oxide all around. Below is shown the band-edge diagram for CB (Ec) and VB(Ev). (b) Electrostatic model based on capacitive coupling used for self-consistent solution in the NWFETs. Cg, Cs and Cd are the gate, source and drain capacitance. Efs/ Efd represent the source/drain Fermi level.

## II. MODELING APPROACH

The electronic structure of SiGe NW channel is calculated using a Virtual Crystal Approximation (VCA) method where each atom is approximated as a fictitious SiGe atom whose Tight Binding (TB) parameters are the weighted average of the individual TB parameters of Si and Ge [5]. A 20 band sp$^3$d$^5$s$^*$ TB model with spin orbit (SO) coupling is used in the electronic structure calculation [5-8]. The TB-VCA model accurately reproduces the experimentally verified bulk bandstructure of relaxed and strained SiGe systems and is well suited to handle material and strain variations at the atomic level [6].

The charge and the potential in SiGe NWFETs are obtained by self-consistently coupling the electronic structure calculation to a 2D Finite Element (FEM) Poisson solver. The FET electrostatics is represented by a capacitive model (Fig. 1(b)), where the channel potential is controlled by the source (S), gate (G) and drain (D) capacitance. In our calculations we assume $C_s = C_d = 0$ (full gate control). After obtaining the self-consistent charge and potential the terminal characteristics are obtained using a ballistic top of the barrier (ToB) transport model whose details are provided in Ref. [6-8] and the references therein.

*Device details:* In this work circular NWFETs with a 9 nm core diameter (CD) and <100> and <110> channel orientations are considered. Ge concentrations of 50%, 70% and 90% are used. The gate oxide thickness (T$_{OX}$) is set to 1.5nm with a relative dielectric constant ($\varepsilon_r$) of 3.9. The gate bias (V$_{GS}$) is swept from 0 to 1V and the drain bias (V$_{DS}$) is fixed at 0.5V for self-consistent simulations. Source and drain are n(p)-doped for the n(p) FETs while the channel is assumed intrinsic.

978-1-4244-6572-9/10 $26.00 © 2010 IEEE

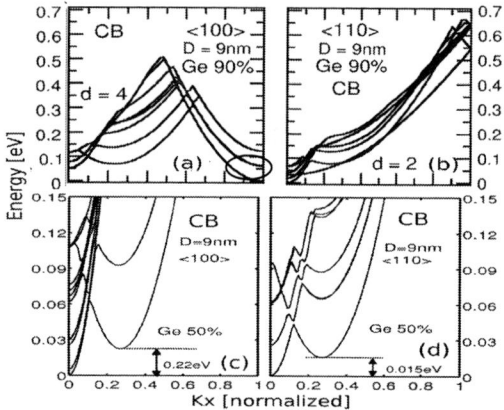

**Figure 2.** Conduction Band (CB) dispersion for circular SiGe NW with 9nm diameter. CB of 90% Ge NW with (a) <100> and (b) <110> channel. CB of 50 % Ge NW with (c) <100> and (d) <110> channel. Band minima for all the wires adjusted to 0eV for better comparison.

## III. RESULTS AND DISCUSSION

### A. Electronic Dispersion in SiGe NWFETs

The bandstructure is an important quantity to calculate the transport properties of ultra-scaled devices. Fig. 2 and 3 illustrate the CB and the VB respectively for <100> and <110> oriented NW with 90% and 50% Ge concentrations. Only half of the k space is shown since the +k and –k dispersions are symmetric. All the band minima (CB) and maxima (VB) have been shifted to 0eV for better comparison.

The number of sub-bands, their relative position in energy and the position of the CB minima are strongly governed by the Ge concentration and the NW channel orientation. For <100> channels at Ge >= 90%, the CB minima appears at the Brillouin edge due to the projection of the bulk L-valley of Ge (Fig. 2a).In the case of 50% Ge, the CB minima is at Γ due to the projection of the $\Delta_4$ valleys of Si (Fig. 2c) . For <110> wires the CB minima is always at Γ (Fig. 2b-d).

The CB minima valley degeneracy is 4(2) for <100>(<110>) wires for 90% Ge(Fig. 2a-b). This difference is due to the projection of different bulk CB valleys. The valley separation is considerably different for different Ge% as well as channel orientations as shown in Fig. 2c-d.

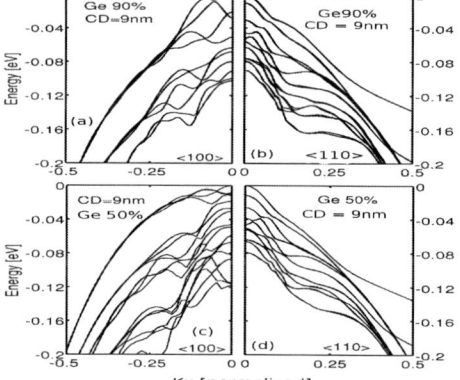

**Figure 3.** Valence Band (VB) dispersion shown for SiGe NW with CD = 9nm. VB of 90% Ge NW with (a) <100> and (b) <110> channel. VB of 50 % Ge NW with (c) <100> and (d) <110> channel. Band maxima for all the wires adjusted to 0eV for better comparison. VB E k is heavily warped.

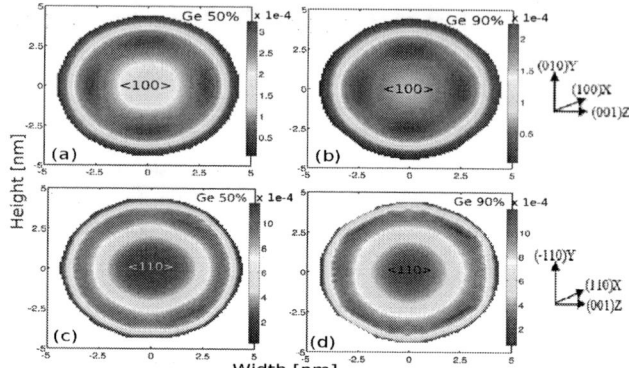

**Figure 4.** Inversion electron density distribution in the cross-section plane of the SiGe n-NWFETs. Wire diameter is 9nm.$V_{GS}$ = 0.8V, $V_{DS}$ = 0.5V. <100> oriented channels are shown with (a) 50% Ge content and (b) 90% Ge content. <110> channels shown with (c) 50% Ge and (d) 90% Ge. Charge distribution is strongly governed by Ge% and channel orientation.

Fig. 3 shows the VB E-k for <100> and <110> oriented NWs for Ge 90% and 50% which are heavily warped. This makes a full band treatment of the VB very important to capture this anisotropy of the E-k. The overall VB E-k looks quite similar for different Ge% and orientations due to the similarity of the VB dispersions in Si and Ge [5]. However, the detailed sub-band energy positions are different (Fig. 3).

### B. Charge Distribution in SiGe NWFETs

The total charge in a 1D conductor is given by (1) [9,10], where $|\psi|^2$ is the in-plane carrier probability distribution, f is the Fermi-Dirac distribution function, n and k are the number of sub-bands and k points respectively. $E_{f1,2}$ are the source and drain Fermi-levels respectively. The in-plane spatial charge distribution is governed by the in-plane local density of states (LDOS ~ $|\psi|^2$ ) which can be obtained by the in-plane quantization masses ($m_1$ and $m_2$) [9,10].

$$\rho = 2\Sigma_{n,k}|\psi_{n,k}|^2\Big[f(E_{f1}) + f(E_{f2})\Big] \propto \sqrt{m_1 \cdot m_2} \quad (1)$$

The 1D charge distribution in Fig. 4 and 5 are shown on a single slice at the 'top of the barrier' in the nanowire channel. All the distributions are obtained at $|V_{GS}|$ = 0.8V and $|V_{DS}|$ = 0.5V.

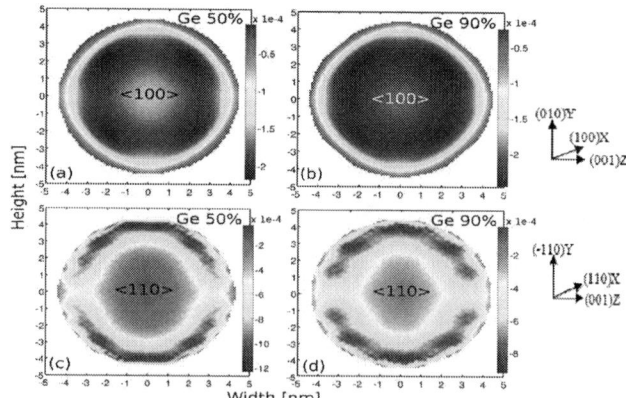

**Figure 5.** Dependence on Ge concentration and channel orientation of the inversion hole density in SiGe p-NWFETs. Wire diameter is 9nm. $|V_{GS}|$ = 0.8V and $|V_{DS}|$ = 0.5V. <100> oriented NWFET shown with (a) 50% Ge content and (b) 90% Ge content. <110> oriented NWFET shown with (c) 50% Ge content and (d) 90% Ge content. In <100> wires charge is pushed along the periphery whereas in <110> wires charge is pushed to the (-110) surface due to the higher in plane DOS in that direction [8].

978-1-4244-6572-9/10 $26.00 © 2010 IEEE

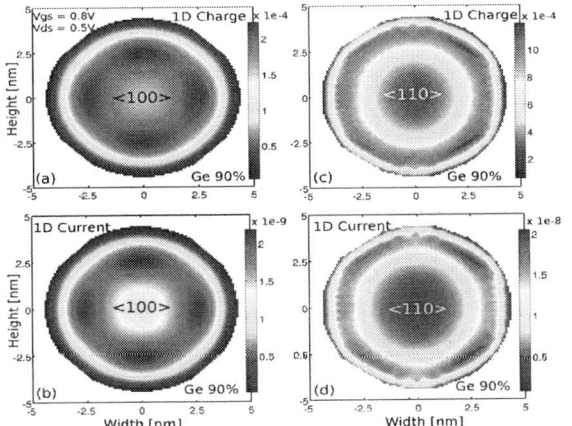

**Figure 6.** 1D Current (A/nm) and charge (#/nm) distribution in 9nm diameter SiGe n-NWFETs. The distribution is at $V_{GS}$ = 0.8V and $V_{DS}$ = 0.5V. (a) Current and (b) charge distribution for <100> NWs. (c) Current and (d) charge distribution for <110> NWs. Charge distribution is mainly governed by the in-plane effective mass ($m^*_{DOS}$) whereas the current distribution also depends on the effective transport mass ($m^*_{tr}$). (2)

The transport (X) and the confinement directions (Y and Z) are shown in the inset of Fig. 4 and 5. The unit of 1D charge density is #/nm.

For the <100> oriented n-FETs electrons show a preferential movement towards Y and Z directions for 50% Ge (Fig. 4a). This is due to the higher quantization masses ($m_{1y} = m_t^†$, $m_{2y} = m_l^†$, $m_{1z} = m_l$, $m_{2z} = m_t$) along the Y and Z valleys (more Si nature) [9]. In Ge, quantization mass is higher along the diagonal of Y-Z plane [9]. This is reflected by the almost circular charge distribution for 90% Ge (Fig. 4b). The <110> oriented wires have higher in-plane DOS along the diagonal of Y-Z plane for both Si and Ge. This results in symmetrical charge distribution for both 50% and 90% Ge concentration due to the projection of X and L valleys along the confinement directions (Fig. 4c-d).

Hole distribution for SiGe wires with 50 and 90% Ge are

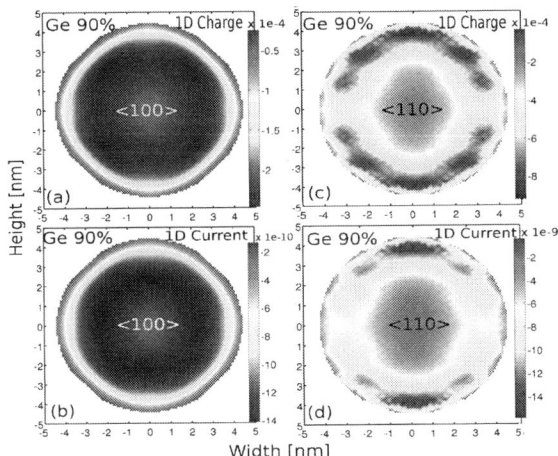

**Figure 7.** 1D Current (A/nm) and charge (#/nm) distribution in 9nm diameter SiGe p-NWFETs. The distribution is at $|V_{GS}|$ = 0.8V and $|V_{DS}|$ = 0.5V. (a) Current and (b) charge distribution for <100> NWs. (c) Current and (d) charge distribution for <110> NWs. <100> wires have isotropic charge and current distribution since the in-plane DOS is isotropic. However, <110> wires show more movement towards (-110) direction due to the higher in-plane DOS along that direction [8].

†$m_l$ = longitudinal mass , $m_t$ = transverse mass

very similar for both <100> (Fig. 5a-b) and <110> orientations (Fig. 5c-d). This happens since the VB dispersion is very similar in Si and Ge and hence in their alloy (Fig. 3). The hole distribution is symmetrically directed towards (110) direction in <100> wires due to the anisotropy and warping in the VB dispersion which affects the in-plane DOS distribution [8].

*C. Difference in spatial charge and current distribution*

In a 1D conductor the ballistic current is given by the product of total charge density ($\rho$) and its velocity at the 'virtual source' [11] ($v_{inj}$) as shown in (2). For a single parabolic band $v_{inj} \sim 1/\sqrt{m^*_{tr}}$, which implies that current distribution in the plane is governed by the in-plane effective masses ($m_1$ and $m_2$) as well as the transport effective mass ($m^*_{tr}$). The regions of 'higher charge density' with small $m^*_{tr}$ are the regions for 'higher current distribution'.

$$I_{1D} = 2q\sum_{n,k}|\psi_{n,k}|^2\left[f(E_{f1}) - f(E_{f2})\right]v_{inj}^{n,k} \propto \sqrt{(m_1 \cdot m_2)/m^*_{tr}} \quad (2)$$

Fig. 6 and 7 compare the current and charge distribution in $Si_{0.1}Ge_{0.9}$ NWFET. Fig. 7a-b show the charge and current distributions for the <100> oriented n-FET. An important observation is that current concentration reduces drastically at the center compared to the edges of the NW. This happens since the $m^*_{tr}$ at the center is dominated by the heavy $m_l$ of Ge ($m_l^{bulk,Ge} \sim 1.5m_0$) which is projected from the Ge L-valley (see the CB E-k in Fig. 2a) where as the $m^*_{tr}$ at the edges is dominated by the much lighter $m_t$ of Ge ($m_t^{bulk,Ge} \sim 0.08m_0$). For the <110> oriented n-FET both the charge and the current distributions are quite similar (Fig. 6c-d). The reason is that regions with higher in-plane DOS mass (high electron density) also have smaller $m^*_{tr}$ ($m_{[110]}^{bulk,Ge,Si} \sim 0.3m_0$).

In the p-type NWFETs there is not much difference in the hole charge and current distribution for <100> and <110> orientations (Fig. 7). This happens since in p-FETs the regions with higher in-plane DOS masses also have smaller transport mass ($m^*_{tr}$) [8]. Holes have the advantage that regions with higher DOS also have higher velocity providing more improvement compared to the electrons where all the higher in-plane DOS regions may not have higher velocity.

*D. Transfer characteristics and performance comparison*

The transfer characteristics ($I_D - V_{GS}$) have been

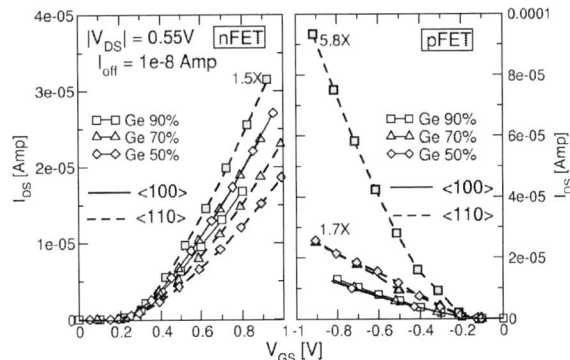

**Figure 8.** Terminal characteristics for <100> and <110> 9nm SiGe NWFETs for 50, 70 and 90% Ge. Left (Right) panel shows the n-FETs (p-FETs). $I_{OFF}$ has been matched to 1e-8A for all the FETs. All the comparisons of $I_{ON}$ for <110> FETs are with respect to the corresponding <100> SiGe FET.

978-1-4244-6572-9/10 $26.00 © 2010 IEEE

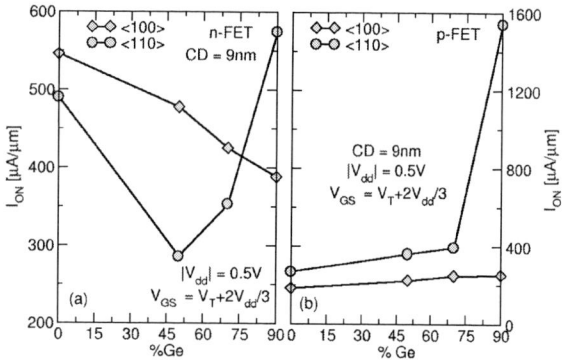

**Figure 9.** On state ballistic current ($I_{ON}$) distribution in 9nm diameter SiGe (a) n and (b) p-NWFETs for different Ge concentrations. For Ge > 80% <110> wires show higher $I_{ON}$ compared to the <100> oriented SiGe n- and p-FETs.

obtained for 9nm diameter ballistic n and p-NWFETs with 50%, 70% and 90% Ge concentration. All the $I_D$-$V_{GS}$ plots have the same $I_{OFF}$ of 1e-8 Amp (Fig. 8). For <100> n-FETs the drain current ($I_D$) reduces with increasing Ge%, whereas for <110> n-FETs $I_D$ increases with increasing Ge% (Fig. 8). This behavior is the result of the higher carrier velocity at the virtual source ($v_{inj}$) in the CB for <100> Si compared to <100> Ge [12]. Hence, adding more Ge reduces the $I_D$ in <100> wires. For <110> direction Ge has higher $v_{inj}$ compared to Si which increases $I_D$ [12].

In p-FETs the <110> orientation is better for hole transport compared to <100> direction due to lighter transport mass and higher DOS [12]. Also <110> Ge shows higher ballistic $I_D$ compared to Si [12]. Due to these reasons <110> p-FETs show higher $I_D$ compared to <100> wires which improves with increasing Ge% (Fig. 8b).

The ballistic $I_{ON}$ for <100> n-FETs degrades with increasing Ge%. In the <110> n-FETs increasing Ge% improves $I_{ON}$ after an initial dip compared to Si, which stems from the reduction in DOS (reduction in charge density) after alloying (Fig. 9a). Eventually $v_{inj}$ increases counteracting the charge reduction and improving the $I_{ON}(2)$. For 90% Ge, <110> wire shows ~1.5X more $I_{ON}$ compared to the <100> wire. In p-FETs $I_{ON}$ improves for both <100> and <110> with increasing Ge%. For 90% Ge, <110> p-FET shows ~5.8X more $I_{ON}$ compared to the <100> p-FET due to very light $m^*_{tr}$ (Fig. 9b).

Intrinsic gate delay ($\tau_D = dC_G/dI_D$) is an important metric for comparing FETs at the circuit level [6]. Smaller $\tau_D$ is desirable for faster circuit operations. For n-FET devices $\tau_D$ decreases with increasing Ge% for <110> direction (Fig. 10a). For p-FETs both <100> and <110> show smaller gate delay with increasing Ge %. Improvements are much bigger for <110> p-FETs (Fig. 10b).

## IV.    CONCLUSIONS

A general tight-binding based VCA method has been presented for obtaining the electronic structure and the terminal characteristics of ballistic SiGe NWFETs. The CB dispersion strongly depends on the Ge concentration in NW whereas the VB dispersions are quite similar for different Ge% studied in this work. The spatial distribution of electron and hole 1D charge and current density has been presented and explained in terms of the in-plane DOS mass ($m_1$, $m_2$) and the transport mass ($m^*_{tr}$). These distributions are strongly influenced by the Ge concentration and the channel orientation. The terminal characteristics and the performance metrics for 9nm circular ballistic SiGe n- and p-NWFETs are provided. SiGe NWFETs with Ge > 80% and <110> orientation are found the most suitable candidate for replacing Si for both n and p type FETs due to their higher $I_{ON}$ and smaller gate delays ($\tau_D$) compared to Si.

### ACKNOWLEDGMENT

Financial support for this work provided by SRC, MSD and NSF. Authors also acknowledge the computational resources provided by nanoHUB.org, a project of NCN.

### REFERENCES

[1] ITRS Technology Roadmap, url: http://www.itrs.net/reports.html
[2] Y. Jiang, N. Singh, T. Low, P. Lim, S. Tripathy, G. Lo, D.Chan, and D.-L.Kwong, "Omega-gate pMOSFET with nanowire like SiGe/Si Core/Shell Channel, IEEE EDL, 30, 4, pp. 392-394, (2009).
[3] Y. Jiang, N. Singh, T. Low, W. Loh, S. Balakumar, K. Hoe, C. Tung, V. Bliznetsov, S. Rustagi, G. Lo, D. Chan, and D. Kwong,"Ge-rich (70%) SiGe nanowire MOSFET fabricated using pattern-dependent Ge-condensationtechnique." IEEE EDL, 29, 6, pp. 595-598, (2008)
[4] Z. Cheng, J. Jung, M.L. Lee, H. Nayfeh, A.J. Pitera, J.L. Hoyt, E.A. Fitzgerald, and D.A. Antoniadis, "SiGe-On-Insulator(SGOI): two structures for CMOS application," Advanced Materials for Micro-and Nano-Systems (AMMNS), Tech.Rep., (2003).
[5] T.B Boykin, G. Klimeck and F. Oyafuso, "Valence band effective mass expressions in the sp3d5s' empirical tight-binding model applied to a Si and Ge parameterization," PRB, 69, pp. 115201, (2004).
[6] A. Paul, S. Mehrotra, M. Luisier and G. Klimeck, "Performance prediction of ultra-scaled SiGe/Si core/shell electron and hole nanowire MOSFETs", to appear in IEEE EDL, (2010).
[7] N.Neophytou, A.Paul, G.Klimeck,and M.Lundstrom, "Bandstructure Effects in silicon nanowire electron transport," IEEE TED, 55, pp. 1286-1297, (2008).
[8] N. Neophytou, A. Paul, and G. Klimeck, "Bandstructure effects in silicon nanowire hole transport", IEEE Trans. on Nanotech., Vol. 7, pp. 710 - 719 (2008).
[9] F. Stern and W.E. Howard, "Properties of semiconductor surface inversion layers in the elecric quantum limit", Phys. Rev. 163, pp. 816-835, (1967).
[10] M. Bescond, N. Cavassilas, K. Kalna, K. Nehari, J.L. Autran, L. Raymond, M. Lanoo and A. Asenov, "Ballistic transport in Si, Ge and GaAs nanowire MOSFETs", IEEE IEDM, (2005).
[11] A.Khakifirooz and D.Antoniadis, "Transistor performance scaling: The role of virtual source velocity and its mobility dependence," in IEDM, doi:10.1109/IEDM.2006.346873, (2006).
[12] J. Wang, A. Rahman, G. Klimeck, and M. Lundstrom, "Bandstructure and orientation effects in ballistic Si and Ge nanowire FETs," in IEDM, doi:10.1109/IEDM.2005.1609399, (2005).

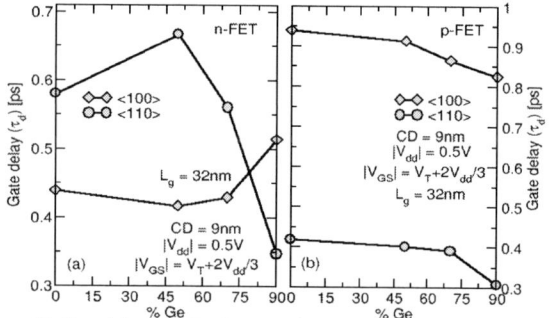

**Figure 10.** Gate delay ($\tau_D$) distribution in 9nm diameter SiGe (a) n and (b) p-NWETs for both <100> and <110> oriented channels. Assumed gate length (Lg) of the device is 32nm. <110> p-FETs are ~60% faster than <100> pFETs for all Ge compositions.

# Continuous-Time/Discrete-Time (CT/DT) Cascaded Sigma-Delta Modulator for High Resolution and Wideband Applications

Ali Mesgarani

*Electrical and Computer Engineering*
*University of Idaho*
*Moscow, Idaho, USA*
mesg9914@vandals.uidaho.edu

Khosrow H.Sadeghi

*Electrical Engineering*
*Sharif University of Technology*
*Tehran, IRAN*
ksadeghi@sharif.edu

Suat U. Ay

*Electrical and Computer Engineering*
*University of Idaho*
*Moscow, Idaho, USA*
suatay@uidaho.edu

*Abstract*—**This paper reports transistor-level design of a new continuous-time (CT), discrete-time (DT) cascaded sigma delta modulator (SDM). The combination of a CT first stage and a DT second stage was utilized to realize a high speed, high resolution analog-to-digital converter (ADC). Power consumption of CT first stage is lowered by optimizing the gain coefficients of CT integrators in a feedforward topology. Moreover double sampling (CDS) was used in second stage integrators to further reduce power consumption. Proposed new SDM is simulated in 0.18μm CMOS technology and achieves 84dB dynamic range for a 10MHz signal bandwidth. Total analog power dissipation measured was 44mW.**

*Keywords*— **Sigma delta modulator, ADC, continuous-time discrete-time, CTDT modulator, high speed ADC.**

## I. INTRODUCTION

The proliferation of wireless communications drives the demand for high-speed analog-to-digital converters (ADCs) with wide dynamic range, wide bandwidth and low power consumption. Wideband sigma delta ADCs with low over-sampling ratios (OSR) are currently being investigated for such applications [1, 2]. The resolution of discrete-time (DT) sigma delta modulators (SDM) is limited by settling accuracy of switched capacitor integrators. Although double sampling technique has been widely used to improve the speed performance of DT SDMs [3], incomplete settling at high sampling frequencies is still an issue in the first DT integrators since they have to settle within the overall resolution of ADC in a clock period. Moreover mismatch between the two sampling paths of first DT integrator is introduced directly to the output of the modulator. Hence DT SDMs are not an alternative for wideband high accuracy applications because of the power limitations. On the other hand since the performance of continuous-time (CT) SDM is not limited by settling accuracy, they can operate at higher sampling frequencies. Sampling frequencies of 640MHz and 950MHz have been reported in the literature [2], [4]. Although CTSDMs can operate at very high sampling frequencies, they are more sensitive to circuit nonidealities such as clock jitter,

mismatch between elements of the feedback digital to analog converter in multi-bit SDMs, integrator time constant variations and excess loop delay that limit the resolution of the modulator [5].

This paper reports design and circuit implementation of a CT/DT cascaded SDM and proposes ideas at system and circuit level to reduce the power dissipation. Section II describes the new architecture, and design of the modulator blocks. In section III some of the most important nonideal effects on the modulator have been investigated. Circuit implementation of the modulator is briefly described in section IV and the simulation results in HSpice are presented in section V. Conclusions are given in section IV.

## II. CT/DT CASCADED SDM ARCHITECTURE

The modulator in this paper is designed to have effective resolution of 13bits within 10 MHz signal bandwidth. Dynamic range (DR) of an $L^{th}$ order SDM with an N-bit internal quantizer is given in (1).

$$DR = \frac{3}{2}\frac{2L+1}{\pi^{2L}}OSR^{2L+1}(2^N-1)^2 \quad (1)$$

Different combinations of modulator order (L), number of bits of internal quantizer (N) and oversampling ratio (OSR) can be used to meet the DR specification. However, important limitations preclude choosing some of the values. They are;

- Increasing the OSR leads to increased sampling frequencies which further increases power dissipation.

- High order single loop SDMs may suffer from instability.

- Small values for N may increase jitter induced error [6], while large values for N increases the area and power.

Taking these in consideration and after investigating different combinations of modulator parameters in MATLAB, a 2-2 cascaded SDM with four bit quantizers in both stages and an OSR of 10 was selected for this design. By choosing these values the jitter error is reduced effectively and the quantization noise remains well below the expected resolution.

## A. Feedback versus Feedforward Topology in SDM

Fig. 1(a) shows a block diagram of a second-order feedback SD ADC with 4bit quantizer. Assuming a linear model, the signal transfer function (STF) is given as $z^{-2}$ and quantization noise transfer function (NTF) is equal to $(1-z^{-1})^2$. The STF is simply a two-period delay. Then, signal input to the loop is given by:

$$X - Y = (1-z^{-2})X - (1-z^{-1})^2 E_Q \qquad (2)$$

where X and Y are input and output of the SDM and $E_Q$ is the quantization noise. Hence, the input of loop filter includes not only the shaped quantization noise, but also the signal component. This component causes the large signal swing at the output of each integrator leading to nonlinear effects in the integrator amplifier.

Fig. 1(b) shows a second-order feedforward SD ADC which has the same NTF as the feedback version but a unity signal transfer function. With unity STF, the loop sees only the shaped quantization noise which causes the very small signal swing at each integrator output. The small signal swings make the ADC more robust against opamp nonlinearities. Hence a feedforward topology was selected to implement first stage of the cascaded architecture. The input at the second stage of the modulator only contains quantization noise from the first stage leading to low integrator swings. Hence feedback topology was preferred for the second stage.

## B. First Stage of the Cascaded SDM Design

The first stage of the proposed cascaded architecture is implemented in CT. A CT first stage is preferred over a DT one in terms of power dissipation since the first two integrators are the main power consumers. In order to reduce the power dissipation of CT modulator, gain coefficients of the CT integrators must be chosen carefully [7]. Required bandwidth of amplifiers is reduced considerably by choosing smaller integrator gain coefficients. But reducing them causes some serious limitations, especially in the first CT integrator:

- If the gain coefficient of the first CT integrator is chosen too small, input referred noise from the second integrator is not attenuated considerably. Hence, a high performance amplifier in the second integrator is required, which increases overall power dissipation of the modulator.

- Choosing a small value for the first integrator gain coefficient leads to large integrating capacitor that forces large currents at the output of the amplifier.

- A large integrating capacitor imposes large DAC currents which further increase the DAC power dissipation.

Considering above limitations and the distribution of integrators gain coefficients, gain coefficient for the first CT integrator was chosen as small as 1/2 to provide enough attenuation on the input referred noise of the second integrator. Gain coefficient of the second integrator was chosen 1/4. Values less than 1/4 increase the input referred thermal noise. Notice that these integrator coefficients are only practical in a feedforward topology with three degrees of freedom in forward loop gain, since a forward loop gain less than unity increases the input referred quantization noise.

Different approaches have been proposed to design CT modulators. Direct design in CT domain [1] and DT to CT transformation methods [5] have been widely used. In this design second approach was used to map DT loop filter shown in Fig. 1 (a) on the equivalent CT modulator. Applying the transformation on the DT loop filter shown in Fig.1(b) we get:

$$H(s) = L^* \left( \frac{2z^{-1} - z^{-2}}{(1-z^{-1})^2} \right) = \frac{1.5}{sT} + \frac{1}{(sT)^2} \qquad (3)$$

Where $L^*$ is the starred Laplace transformation. The only problem with implementing the loop filter using feedforward architecture is the full scale swing at the output of the summing amplifier. To accommodate this problem, the loop filter is multiplied by 1/2 and a gain of 2 is introduced to the quantizer. In this way the signal swing at the output of the adder output reduces to half of the input full scale. The multiplication by 2X in the quantizer is easily implemented by comparing the summing amplifier output to half of reference voltages used in the quantizer. Fig. 2 shows the implementation of CT loop filter using a feedforward topology.

## C. Second Stage of the Cascaded SDM Design

In a CTSDM the signal transfer function cannot be demonstrated fully in DT domain [8]. Different approaches have been proposed in design of cascaded continuous time sigma-delta modulators. In one case, bilinear transformation is used to realize the noise cancelation logic, [9]. This approach is not very accurate and causes the quantization noise from the first stage to leak into the overall output degrading Signal to Noise Ratio (SNR). In another case, the state space method which provides complete matching between analog and digital coefficients and completely removes first stage quantization noise from the output was used, [5]. This approach is very complicated using several feedforward paths and also increases thermal noise.

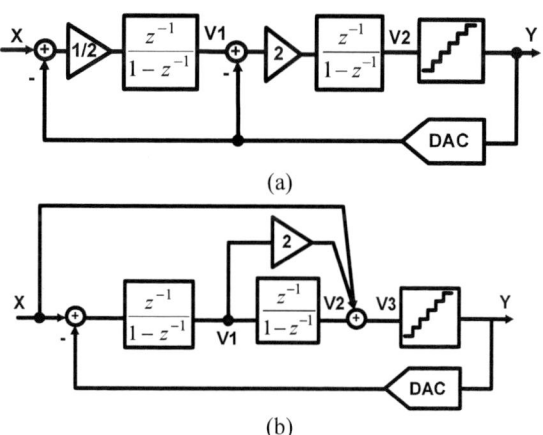

(a)

(b)

Figure 1. A second-order a) feedback b) feedforward SDM topology.

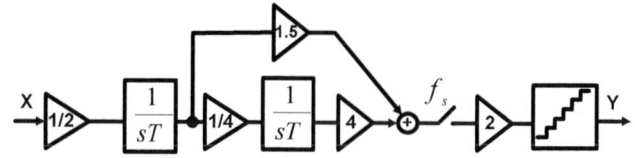

Figure 2. CT loop filter implementation in a feedforward topology

978-1-4244-6572-9/10 $26.00 © 2010 IEEE

A CT second stage requires extra circuits for tuning RC time constant of the CT integrators. Also, a CT second stage suffers from the large mismatch between DAC unit elements which degrades the SNR performance of the modulator in spite of being shaped by second order noise shaping. Use of a DT second stage was first suggested in [7]. In this case, the quantization noise from the first stage can be directly fed into second stage and in ideal case it is completely removed from output using simple digital error cancellation. This makes the design much simpler compared to state space method.

The power requirement on the second stage integrators is much less than the first stage because of the noise shaping characteristic of SDMs. Moreover, utilizing double sampling integrators in the DT second stage power consumption of integrators is reduced significantly making it more comparable to CT ones. Since double sampling integrators are used in the second stage of the modulator, mismatch between sampling paths is shaped by second order noise shaping which prohibits SNR degradation due to this mismatch. Also since the coefficients of switched capacitor integrators are determined by capacitor ratios no extra tuning circuit is required which further reduces power dissipation. Finally since the matching between switched capacitor DAC elements is about 10 times better than current mode CT DACs, the performance of the modulator is less affected by nonlinearity of second stage DAC. Taking into account above considerations a DT second stage was chosen.

### III. NONIDEAL EFFECTS ON SDM PERFORMANCE

Effects of nonidealities that limit the performance of proposed cascaded modulator have been investigated at system level in this section.

#### A. Integrator Time Constant Variations

The time constant of CT integrators is determined by RC product in active RC integrators. Due to process and temperature variations these values and hence integrator time constant may vary by 50% changing the analog coefficients. SNR degrades because the low order shaped quantization noise form the first stage leaks into the output due to mismatch between digital and analog coefficients. Fig. 3 shows the SNR of the cascaded modulator versus time constant variation of integrators. To prevent the degradation of SNR performance of the proposed modulator, this variation has to remain within ±5% of its ideal value. The RC integrator time constants can be tuned digitally utilizing a 4-bit bank of digitally switched binary weighted capacitor.

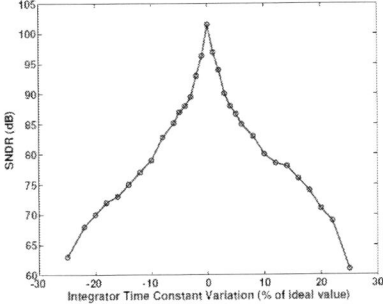

Figure 3. Effect of error in the absolute value of the integrator time constants

#### B. Clock Jitter

Another destructive effect in design of CT modulators is the clock jitter. It can be shown that multi-bit Non Return to Zero (NRZ) DAC waveform can attenuate jitter induced error considerably [6].

#### C. Excess Loop Delay

The building blocks in a modulator loop including the comparator, the DAC linearization logic and the DAC cause delay in the feedback path. This delay increases the order of CTSDM and makes the loop unstable. Different approaches have been suggested for compensation of excess loop delay [10], but only intentional delay insertion removes this problem without the need for RZ feedback waveforms that increase the sensitivity of the modulator to clock jitter. Fig. 4 shows the proposed cascaded modulator with compensation of excess loop delay in CTDSM. The weight of feedforward path from the input to the first stage quantizer was optimized through simulation for maximum SNR. Notice that second stage integrators are clocked at 100MHz while the quantizer is clocked at 200MHz.

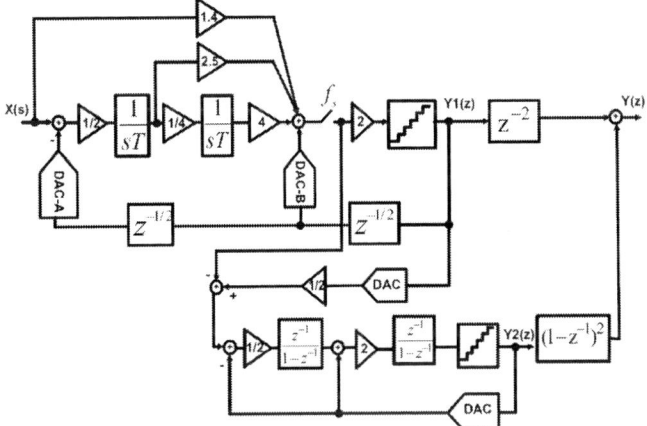

Figure 4. Proposed CT/DT cascaded SDM

### IV. CIRCUIT IMPLEMENTATION OF SDM

The modulator is simulated using 0.18μm CMOS process in HSpice environment and operates from a 1.8V supply. Main building blocks of the modulator were selected and sized according to the system-level specifications. Among the others, the most critical subcircuits are the opamps.

Because of the high linearity requirement, Gm –C circuits were not an alternative for the CT integrators. The CT integrators were realized as active-RC opamps. Two stage operational amplifiers with cascode compensation were used to improve the bandwidth over conventional Miller compensation. Figure5 shows the two stage amplifier used in CT integrators and adder along with common mode feedback circuit. Since the input to the second stage is only the quantization noise from the first stage, the signal swing at the output of integrators is very small. Moreover the load of CDS integrators is only capacitive, hence single stage folded cascade amplifiers were used in the second stage.

The specifications of amplifiers are summarized in Table I.

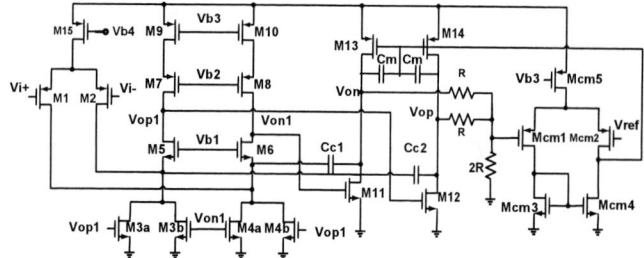

Figure 5. Two stage amplifier for CT integrators and the adder

## V. SIMULATION RESULT

The 2-2 cascaded CT/DT SDM was implemented in 0.18μm technology. Fig 6 shows the output spectrum of the modulator. In this simulation all resistors were intentionally deviated by -4% from their ideal values to consider for the worst case SNR. A full-scale sinusoidal input with frequency of 1MHz was applied at the input. Maximum SNDR of 83dB was achieved. The total analog circuits power dissipation was measured was 44mW. The simulated DR is shown in Fig 7 for 10MHz sine input. The cascaded modulator achievers 84dB dynamic range for a 1MHz sine input.

## VI. CONCLUSION

Design of a 1.8V, 0.18μm CMOS, 84dB DR, 10-MHz CT-DT cascaded SDM has been presented. A CT first stage has the advantage of lower power integrators and inherent anti-aliasing filtering. Although a CT second stage benefits from slower amplifier, a DT second stage was preferred over a CT one, in terms of design simplicity, power dissipation and SNR performance.

TABLE I.    AMPLIFIER SPECIFICATIONS IN STAGES

|  | Gain DC | Unity Gain Bandwidth | Phase Margin | Power |
|---|---|---|---|---|
| First CT integrator | 72dB | 710MHz | 77° | 11.4mW |
| Second CT integrator | 72dB | 370MHz | 58° | 3.8mW |
| Adder | 64dB | 410MHz | 68° | 4.9mW |
| First DT integrator | 56dB | 380MHz | 58° | 4.3mW |
| Second DT integrator | 54dB | 400MHz | 64° | 4.6mW |

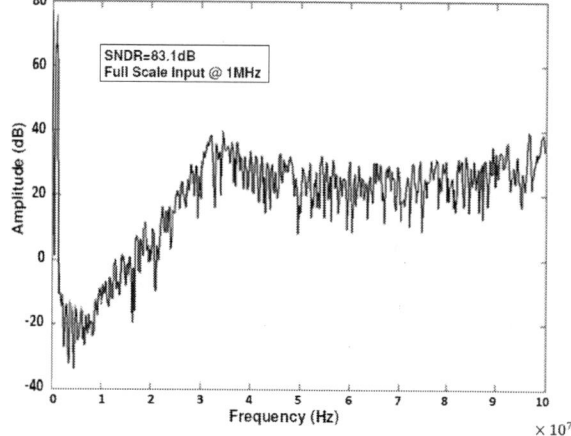

Figure 6. Output spectrum of cascaded SDM for 1MHz full scale sinewave.

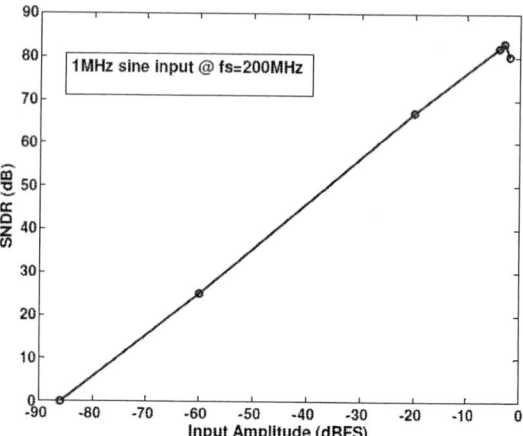

Figure 7. DR of the modulator for a 1MHz sine input

Utilizing a feedforward structure and integrator gain coefficients adjustment, first stage power dissipation can be minimized. Also double-sampling integrators in the second stage of the cascaded architecture help to reduce second stage power dissipation.

## REFERENCES

[1] Ramon Tortosa, Jose Manuel de la Rosa Utrera, Angel Rodriguez Vazquez, Francisco Vidal Fernandez Fernandez , "Design of a 1.2-V cascade continuous-time sigma-delta modulator for broadband telecommunications" , ISCAS 2006, pp. 589-592

[2] G. Mitteregger, et al, "A 20- mW 640- MHz CMOS continuous- time ADC with 20- MHz signal bandwidth, 80- dB dynamic range and 12-bit ENOB," IEEE JSSC, vol. 41, no. 12, pp. 2641- 2649, December 2006.

[3] Tae-Seong Jeong, Wooseok Choi, Jun-Gi, Changsik Yoo, "Low voltage analog to digital converter using sigma-delta modulator" 2008 International SoC Design Conference, III-52-III-53

[4] Matthew Z. Straayer, Michael H. Perrott, "A 12-Bit, 10-MHz bandwidth, continuous-time sigma-delta ADC with a 5-Bit, 950-MS/s VCO-based quantizer" IEEE JSSC, vol. 43, no. 4, pp. 805- 814, April 2008.

[5] F. Gerfers, M. Ortmanns, Continuous-time sigma-delta A/D conversion: fundamentals, performance limits and robust implementations, Springer; 1ˢᵗ edition December 2005.

[6] H.Shamsi, O.Shoaei, R. Doost " Analysis of the clock jitter effects in a time invariant model of continuous time delta sigma modulators ", IEICE Transactions on Fundamentals of Electronics, Communications, and Computer Sciences , vol. E89-A, no.2, pp. 399-407, February 2006.

[7] Kulchycki, R. Trofin, K. Vleugels, B.A.Wooley, "A 77-dB dynamic range, 7.5-MHz hybrid continuous-time/discrete-Time cascaded sigma delta modulator" IEEE JSSC, vol. 43, no. 4, pp. 796-804, April 2008.

[8] R.Schreier, and G. C. Temes, "Understanding delta-sigma data converters" . New York: IEEE Press (2005)

[9] L.J. Breems, R. Rutten, and G. Wetzker, "A cascaded continuous- time modulator with 67-dB dynamic range in 10- MHz bandwidth," IEEE JSSC, vol. 39, no. 12, pp. 2152- 2160, December 2004

[10] M.Keller; A.Buhmann, J.Sauerbrey; M.Ortmanns,Y.Manoli."A comparative study on excess loop delay compensation techniques for continuous time sigma delta modulators," IEEE Tran. on Circuits&Syst., vol.55, no.11,December 2008

# All Digital Multiplying DLL Using Precision Digital Delay Line as DCO

Seong-Hoon Lee
Micron Technology Inc.
Boise, ID 83707, USA
shlee@micron.com

*Abstract*— **In this paper all digital multiplying delay-locked loop (MDLL) is presented, which uses a 3rd order precision digital delay line (HDL) as DCO. Maximum locking frequency at 1.3V was 1GHz with multiplication factor of 50, assuming 20MHz reference clock frequency, and peak to peak jitter was ± 20ps. DCO consumed only 2.2mW, and the rest logic 3.5mW.**

*Keywords-multiplying DLL; DCO; hierarchy delay; phase mixer; jitter*

## I. INTRODUCTION

Though phase-locked loop is a very good solution for high speed clock generation out of lower frequency reference clock signal, it has a couple drawbacks like phase error accumulation and stability issue requiring intense efforts of design. Multiplying DLL (MDLL) architecture has been proposed in recent years as an alternative because of its better jitter performance and superior stability which makes the entire design much easier and more robust [1, 2]. In MDLL, where reference clock periodically refreshes out the jitter accumulation in oscillator, shown in Fig. 2(a).

Thus far, however, all the previous works of MDLL were based on analog VCO requiring charge pump and analog RC loop filters. In this work all digital MDLL with good phase error performance is proposed, which became possible due to a precision digital delay line [3, 4]. Because of all digital design it has many digital advantages like the whole design is much easier, more robust and has high potential to integrate more functions.

## II. Digital delay line and DCO

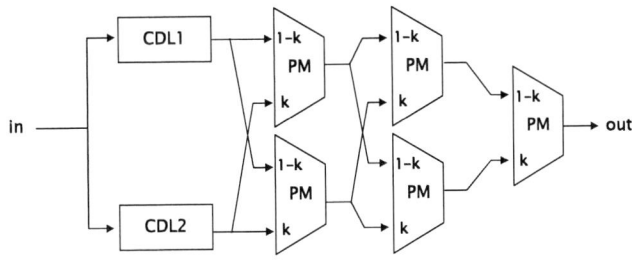

Figure 1.   3rd order HDL.

### A. 3rd order Precision Digital Delay Line

Fig. 1 shows 3rd order HDL idea which will be used as DCO in this work [4]. CDL is coarse delay line composed of logic gates like NAND. PM is phase mixer or phase interpolator and also digital design. Pairs of CDL or PM are used for seamless boundary crossing between delay lines not to allow any glitch in clock signal [3]. CDL provides a coarse step delay time ($t_{CD}$) of 2 logic (NAND) gates and PM provides interpolated phases within $t_{CD}$. If the number of interpolation ($N_I$) per PM was implemented as 10, then 1st stage PM could provide a fine delay time resolution ($t_{FD1}$) of $t_{CD}/N_I$, namely 10ps assuming 100ps of $t_{CD}$, and 2nd stage PM would interpolate again 10 phases out of this $t_{FD1}$ and thus provide finer time resolution $t_{FD2}$ of $t_{FD1}/10 = t_{CD}/10^2$, or 1ps. In the same manner 3rd PM will finally provide a yet finer delay resolution $t_{FD3}$ of $t_{CD}/N_I^3$. If $t_{CD}$ was 100ps then $t_{FD3}$ is 0.1ps on the average. Actual worst case step size is usually bigger than the average value because of non linearity of digital PM [4].

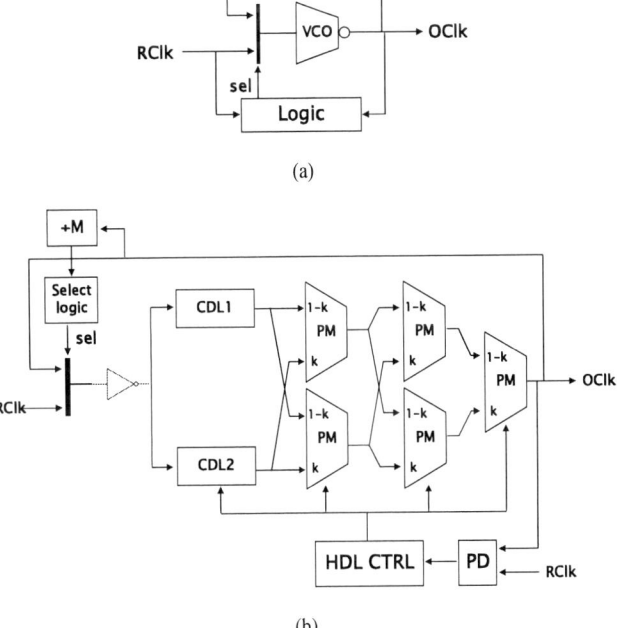

(a)

(b)

Figure 2.   Basic idea of MDLL (a) previous works (b) all digital MDLL using 3rd order DCO of this work.

978-1-4244-6572-9/10 $26.00 © 2010 IEEE

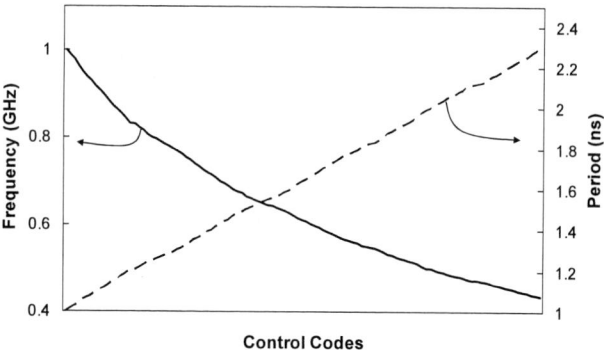

Figure 3. DCO oscillation frequency and period vs. control codes

## B. DCO

Fig. 2 (a) shows a basic idea of previous analog MDLL and (b) proposed digital work [1]. Proposed work is simply done by replacing analog VCO with DCO and collateral control logic. 3rd order HDL composes a DCO including 2 to 1 MUX and inverter. The basic principle of operation is identical to the original one except that everything is done digitally in this work. The DCO frequency is controlled by adjusting the hierarchical delay line. Since the delay time through HDL or oscillation period of DCO is almost linearly varied by control codes, the oscillation frequency is inversely proportional but still monotonic as seen in Fig. 3. Proposed DCO exhibited 1GHz

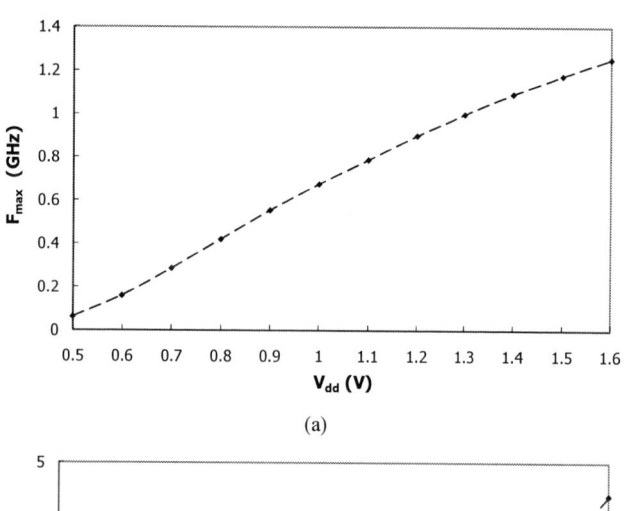

(a)

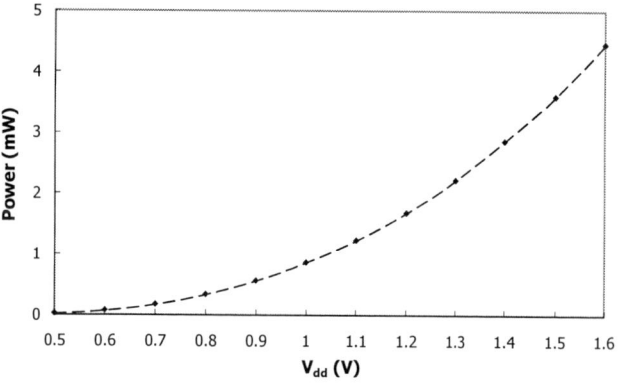

(b)

Figure 4. (a) $F_{OSC,max}$ vs. $V_{dd}$ (b) Power vs. $V_{dd}$ for 3rd order DCO at $F_{OSC,max}$

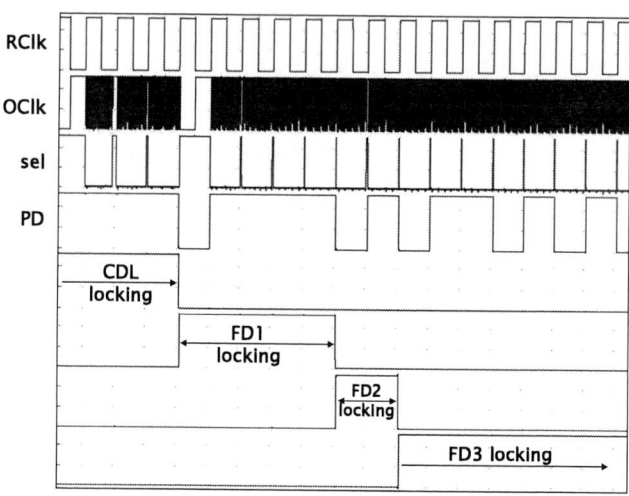

Figure 5. Waveforms for initial locking process

max frequency at 1.3V. Max frequency is limited by the intrinsic delay time ($t_{ID}$) of HDL, which is a sum of propagation delay times of minimum number of logic gates and PMs. The total delay time of HDL ($t_{HDL}$) is sum of $t_{ID}$ and $t_V$ (viable delay time). For N'th order DCO its oscillation period $T_{OSC}$ can be written as

$$T_{OSC} = 2 \cdot t_{HDL} = 2(t_{ID} + t_V) = 2 \cdot t_{ID} + 2 \cdot M(t_{CD}/N_I^N) \qquad (1)$$

M is integer including zero and is controlled by HDL control logic. But it is not generated sequentially but hierarchically because HDL itself has delay hierarchies. Hence max oscillation frequency is $1/(2t_{ID})$ when M=0. 2nd order DCO will have bigger max frequency because it is one PM stage less, but its frequency resolution will be worse. So we have to trade off between max frequency and resolution. Another way to increase the max frequency is to use higher $V_{dd}$. Fig. 4 shows the dependency of max oscillation frequency and power on $V_{dd}$. The frequency increases almost linearly with $V_{dd}$. Minimum frequency is, on the contrary, not limited. Simply increasing M or adding more CDL units will increase $t_{HDL}$, as much. Thus minimum frequency can be lowered more with ease.

### III. ALL DIGITAL MDLL

All digital MDLL was designed and simulated using 3rd order DCO. To achieve fast locking at the beginning, the loop is first controlled using reverse hierarchy at initial locking process. CDL is first adjusted to achieve accelerated locking by adjusting DCO oscillation period with relatively big delay step of $t_{CD}$. After this coarse locking is done 1st PM stage (FD1) is adjusted with finer step size of $t_{FD1}$ or $t_{CD}/N_I$, and then 2nd stage of PM (FD2) is adjusted to further reduce phase error by $t_{CD}/N_I^2$ or $t_{FD2}$. Final step is 3rd stage PM or FD3 locking in this work, and yet finer adjustment is performed with a time resolution of $t_{CD}/N_I^3$ ($t_{FD3}$). Fig. 5 shows this initial fast locking process for M=34 with 20MHz reference clock, i.e. for 640MHz DCO clock signal generation at 1.3V.

The locking process is supposed to be always monotonic so each locking can be noticed when PD's output toggles. Once

978-1-4244-6572-9/10 $26.00 © 2010 IEEE

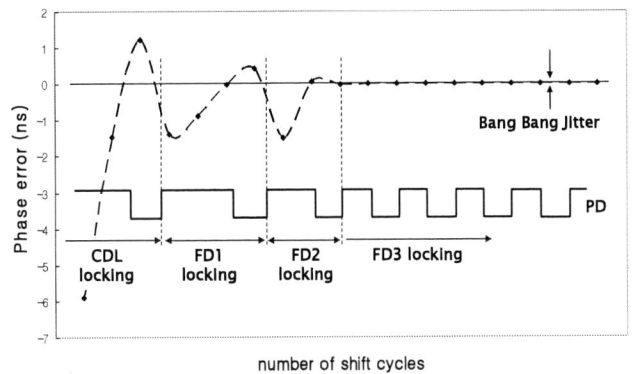

Figure 6. Phase error during initial locking process

initial locking is achieved the DCO is controlled in a correlated manner in order not to allow any glitch in output clock signal. Lower order delay lines are not independently adjusted and dual delay line architecture is fully utilized for this purpose [3]. It takes only tens of reference clock cycles to lock due to above described accelerated locking technique.

Fig. 6 shows how phase error is reduced during the initial locking process. PD output is high when OClk lags behind RClk and low when OClk leads. A transition at PD output means that phase adjustment was too much and requires to be reduced back. Then the loop advances to finer delay adjustment stage. Finally phase locking is achieved but it will toggle back and forth forever since the PD output is binary. The bang-bang jitter was $\pm10$ps for 640MHz (with M=34) case and $\pm20$ps for 1GHz case (with M=50). To suppress the bang-bang jitter after precision phase locking is acquired, a tri-state PD can be used. With a tri-state PD, once the phase error becomes within tolerable range the loop will not toggle the delay line any more. It's easy to construct a tri-state PD out of binary ones. After initial locking is done, phase adjustment is always done with the finest delay resolution only.

## IV. CONCLUSION

All digital multiplying DLL was presented which uses precision digital delay line as DCO. It showed $\pm20$ps phase error for 1GHz oscillation frequency without tri-state PD. Total power was less than 6mW among which 2.2mW was for DCO. Jitter performance can be further improved by using tri-state PD or by increasing $N_I$ more than 10. Using higher than $3^{rd}$ order DCO can also improve the jitter performance but it will decrease $F_{OSC,max}$ as well because $t_{ID}$ increases also. Since digital delay line or DCO is very sensitive to power supply noise, it is highly preferable to put delay lines in a dedicated or regulated power rails.

## REFERENCES

[1] R. Farjad-Rad et al., "A low-power multiplying DLL for low jitter multigigahertz clock generation in highly integrated digital chips," J. Solid-State Circuits, vol. 37, pp. 1804-1812, Dec. 2002.

[2] R. Farjad-Rad et al., "A 33-mW 8-Gb/s CMOS clock multiplier and CDR for highly integrated I/Os", J. Solid-State Circuits, vol. 39, pp. 1553-1561, Sep. 2004.

[3] Jong-Tae Kwak, Chang-Ki Kwon, Kwan-Weon Kim, Seong-Hoon Lee, Joong-Sik Kih, "A low cost high performance register-controlled digital DLL for 1Gbps x32 DDR SDRAM", Symp. VLSI Circuits Dig. Tech. Papers, 2003.

[4] Seong-Hoon Lee, "Variable delay line with multiple hierarchy", U.S. Patent 7 274 236, September 25, 2007.

# Main Memory with Proximity Communication

## A Wide I/O DRAM Architecture

Qawi Harvard and R. Jacob Baker
Department of Electrical and Computer Engineering
Boise State University
Boise, ID, U.S.A.

Robert Drost
VLSI Research Group
Sun Microsystems Laboratory
Menlo Park, CA, U.S.A.

*Abstract* — **The bandwidth and power consumption of dynamic random access memory (DRAM), used as the main memory of a computer system, impacts computer execution rates. DRAM manufacturers focus on density increases, due to the innate price per bit decline of main memory, while processor manufacturers continually focus on boosting performance. This leads to a performance gap between the two technologies. Proximity communication promises to increase the off/on chip bandwidth of DRAM products while reducing the power consumption of the main memory system. The design of a memory system employing 4 Gb DRAM chips with a 64-bit wide communication bus using proximity communication is proposed. Technological roadblocks are analyzed and novel solutions are proposed. The proposed 4 Gb DRAM architecture can reduce the power consumption of a main memory system by 50% while increasing the bandwidth by 100%. The 4 Gb chip architecture measures 68.88 mm$^2$ and has an array efficiency of 59.9%. The estimates are comparable to 2012 International Technology Roadmap for Semiconductors' (ITRS) estimates of 74 mm$^2$ and 56%, respectively.**

*Keywords – DRAM, proximity communication, chip-to-chip, server memory, main memory, bandwidth, power consumption.*

## I. INTRODUCTION

The performance gap between the computer's processor and its main memory has been growing over the past two decades [1]. Density and die size are the figures of merit for main memory manufacturers. Increasing these performance measurements places a physical limit on the latency of the main memory array due to the parasitics [2]. The limitations keep memory latency scaling at roughly 7%, while processor performance has been scaling at roughly 50%. This performance differential is termed the "memory gap", and refers to the growing performance disparity between the processor core and its main memory.

Processor manufacturers have made several architecture changes that enable computer performance to scale with Moore's Law (double the performance every two years). Multiple cores, increased cache levels, multiple threads, and speculative accessing, have made memory stalls almost transparent to the computer user [3]. Main memory manufacturers increase their density per unit area by developing longer bitlines, longer wordlines, decreased unit cell size, and feature size scaling [4]. Main memory manufacturers alleviate bandwidth limitations by using DRAM pre-fetch. Unfortunately, the pre-fetch architectures did not

begin taking hold until 2000 [5]. This places memory bandwidth scaling decades behind processor bandwidth scaling.

Proximity communication is an input/output (I/O) technology that uses capacitors to electrically connect two chips [6]. The off/on chip communication technique has the ability to substantially increase the memory bandwidth and not impact the power consumption [7]. This work develops a memory architecture that utilizes proximity communication to substantially increase bandwidth, while reducing power consumption. This is achieved by allowing a single DRAM chip to provide a full cache line of memory (64 Bytes).

## II. PROXIMITY COMMUNICATION

Capacitive coupled proximity communication is a chip-to-chip interface technology that uses the top level of metal on an integrated circuit to form the parallel plates of a capacitor. Two chips are placed face to face and their top level of metal is allowed to come within close proximity (1 μm – 20 μm) of each other without touching. This arrangement creates a parallel plate capacitor.

### A. Advantages

The advantages of proximity communication allow for a significant reduction of parasitics in the transmission channel, which increases bandwidth and lowers power relative to other chip-to-chip interconnects. Fig. 1 depicts a cross sectional view of two chips using proximity communication as the I/O interface.

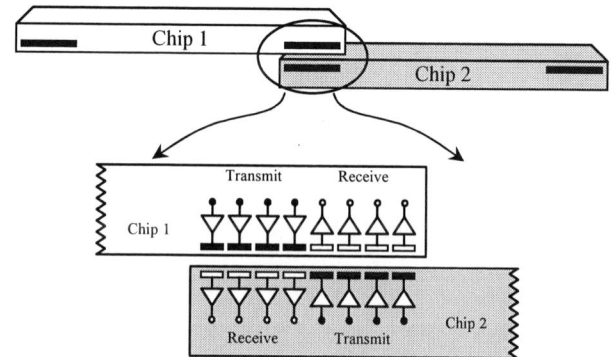

Figure 1. Cross section view of placing two chips face-to-face and within close proximity of each other [6].

The removal of off chip wires allows for a fixed impedance to be delivered to the transmission channel. The passivation over the metal pads is not opened, as in wire-bonded applications, which allows the electrostatic discharge (ESD) protection circuitry to be removed, and for this reason the on-die resistive termination is superfluous [8].

The increase in I/O density is another advantage of proximity communication. The capacitance of parallel plate capacitors is at least 10 pF/mm$^2$ (with a 1 μm gap). 400 I/O channels per mm$^2$ is possible when each communication channel uses 25 fF. The configuration creates a research avenue for scaling the transmission channel below 25 fF.

Placing multiple die into a single package requires complicated wire-bonding technologies used for chip-to-chip interconnects [9]. Proximity communication allows chips to be simply glued in place, which increases the ease of testability. System in package (SiP) configurations can be tested, and defective chips easily replaced while using proximity communication.

### B. Challenges

Chip misalignment is a major challenge associated with the development of proximity communication. Researchers at Sun Microsystems were able to develop a novel solution to this problem [10]. Through the development of electronic sensors, which could be incorporated into the same silicon substrate as transmit and receive circuits, it was possible to determine the misalignment of the two chips. Electrical steering circuits were developed that allowed the transmit data to be driven to multiple receiver pads to electrically realign the transmission channel.

## III. DRAM TRENDS

Incorporated proximity communication into a DRAM architecture without understanding the DRAM market will result in a product that does not meet the need of current memory and computer systems.

### A. Effect of Price Decline and Scaling

The performance differential between microprocessors and the main memory system is referred to as the memory gap and is often misunderstood. The memory gap measures the microprocessor's instructions per second and the main memory's access latency. The confusion occurs when you blindly relate these two figures of merit. DRAM manufacturers focus the majority of their innovations on the process technology that allows for an increase in density. The reason for this is due to the innate price per bit decline of DRAM.

DRAM manufacturers are forced to focus on density scaling over access latency, or I/O bandwidth, due to the historic 36% price per bit decline. Putting this into perspective, if two gigabits of memory chip costs $2.00 today, then four gigabits would cost $1.64 in two years. Density scaling in main memory chips follows Moore's Law. The doubling of the number of transistors in main memory chips every two years (or $\sqrt{2}$ every year) is used to increase the density of the die.

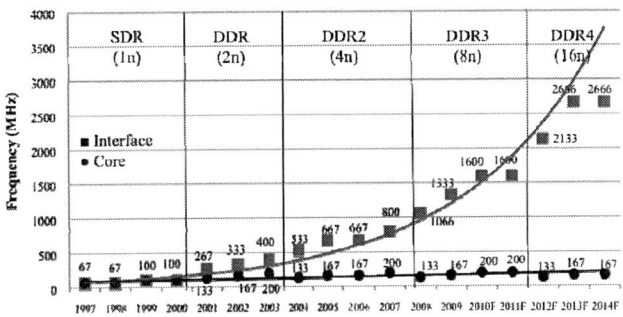

Figure 2. Array pre-fetch, of two and higher, has allowed off chip bandwidth to increase at a rate of 26% per year since 2000, while the core frequency does not scale [5].

Microprocessors use the extra transistors to increase the number of instructions that can be completed each second. Main memory manufacturers have the ability to arbitrarily set the latency and bandwidth of the memory chip. These chips sacrifice power and die size, which creates their inability to compete in the main memory market, which requires large densities [11]. Instead, these products find their place in varying applications that do not require large density.

Reducing the minimum feature size of the components in the memory array achieves the required density scaling. The reduction in feature size increases the parasitics associated with the array and places a physical limit to the bandwidth. The parasitics have placed a limit on the column bandwidth of current memory chips to 133 MHz – 200 MHz. Each generation of main memory starts with a column access of 133 MHz, and then transitions to 167 MHz, and then to 200 MHz to achieve a generational approach to chip bandwidth. Fig. 2 shows this bandwidth trend in main memory chips.

Array pre-fetch allows the DRAM to sustain a larger off-chip bandwidth. Array pre-fetch refers to accessing all bits of the latency at once, and serializing the data in the data path. Main memory chips have been operating with four, eight, or sixteen data pins over the past three generations (DDR, DDR2, DDR3), with eight data pins being most common. The maximum bandwidth that can be achieved with DDR3 chips is 12.8 Gbps (8 data pins, pre-fetch of 8, at 200 MHz). Increasing chip speeds above 12.8 Gbps requires an increase in pre-fetch (to 16) or an increase in data pins, due to the column access limit.

### B. Memory Channel Bandwidth Limitation

Current computer systems use a 64-bit wide data bus to communicate between the main memory and the microprocessor. Series stub terminated logic connections are used in computer system memory channels due to the ease of memory upgrades. The series stub terminated logic refers to terminating electrical signals at each memory module with a resistive pull up device that prevents transmission line reflections from interfering with transmitted data on the shared memory channel.

The resistive termination network, along with module loading, places a bandwidth limit on the memory channel. Server applications require a substantially larger density (or number of main memory modules) then personal computers.

978-1-4244-6572-9/10 $26.00 © 2010 IEEE

Registered, fully buffered, and load reduced DIMMs were developed for server applications to increase the number of DIMMs per memory channel. These innovations have a cost and power premium associated with them.

## IV. X64 DRAM ARCHITECTURE

A 4 Gb DRAM architecture utilizing proximity communication was developed that is realizable with existing technology and meets 2012 ITRS predictions [12]. Challenges associated with incorporating proximity communication into DRAM were characterized and several innovations were developed that alleviated these challenges. A novel global I/O routing structure was discussed that promises to increase the number of data signals that can be read and written to a memory array. The slice architecture was developed to increase the modularity of memory systems.

### A. Moving the Pads

Moving the communication channel to the edge of the DRAM chip creates several interesting challenges when performing an architectural feasibility study. The bank structure used in this research alleviates the initial challenges. Once the communication channel is moved to the edge of the die additional circuitry is required to buffer the signals into the memory chip. Limiting the number of rows per bank creates a "short" bank that reduces global data and command signals, eliminating the need for additional buffers.

The inexpensive process technology of DRAM chips utilizes 2 – 3 layers of metal above the memory capacitor. This places an intrinsic limit to the number of global I/O tracks over each bank. Due to this, the half-bank structure used in this proposal has 64k columns and 8k rows. This half-bank structure must decode the 64k columns into eight 8k pages. A by 64 DRAM chip operating with a pre-fetch of eight requires 512 bits to be accessed at once. Accessing 512 bits from one bank requires the use of a half-bank to reduce the total metal usage. Each half-bank supplies 256 bits of data. This allows the global I/O track to be spread across the chip, limiting metal usage for the global I/O bus. The challenges of buffering the signals into the array and limited routing channels are circumvented by using the proposed bank and segmented page structures. Fig. 3 shows the block diagram of the 4 Gb DRAM die. The half-bank structure can be thought of as dividing each bank horizontally, and firing a wordline in each half-bank.

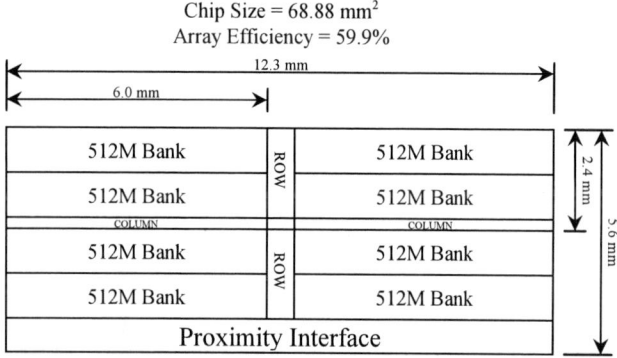

Figure 3. A 4 Gb DRAM architecture incorporating proximity communication and centralized row and column circuitry.

### B. Local I/O Routing

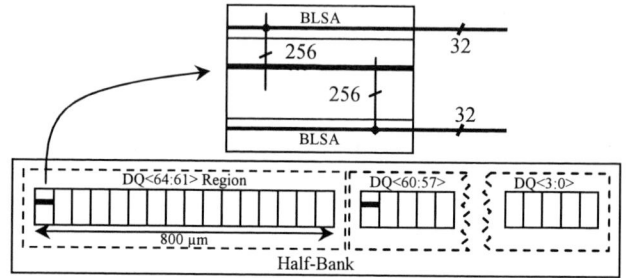

Figure 4. Space and data mapping of the local input/output routing within a half-bank.

The by 16 and by 32 proximity configurations will not require any significant innovation, but the by 64 configuration will require additional innovation for local I/O routing. The large number of global I/O tracks (256 per half-bank) requires 32 data signals from each 256 kb memory array. Moving 32 data signals from the bitline sense amplifiers to the global I/O track is a major challenge due to the limited routing space above the bitline sense amplifiers. Increasing the page size will alleviate this challenge but will also increase the power consumption. Instead, these signals can be routed to the top and bottom of each 256 kb memory segment, as seen in Fig. 4. An additional avenue for architectural research consists of routing the data signals through adjacent inactive bitlines (above and below).

### C. New Global I/O Routing

As mentioned above the memory array operates at a maximum frequency of 200 MHz due to the parasitics of the memory array. The global I/O route does not share the parasitics of the memory array and can operate at a higher frequency. Insertion muxes, and additional latches can be used to keep the global I/O bus fully occupied with data. A column path protocol can be developed that allows for multiple banks to be accessed and data stored in the local I/O channels. Busy, ready, and data insertion requests can be used to allow the global I/O routing to operate at a higher frequency, while the memory array remains operating at frequencies below 200 MHz.

### D. Modular Architecture

Main memory DRAM chips use a large number of repeated structures and symmetry. The proposed modular architecture speeds up design verification. Each modular architecture contains all circuitry required for one data pin to read and write. Combining many of these modular structures together will create the entire chip. A data, command, and clock modular architecture was developed during this research.

The first advantage of this architecture is that the time required for chip verification can be reduced significantly. Due to the sheer number of transistors on a modern DRAM chip, simulating an extracted netlist can take several weeks to complete. Using smaller modular blocks to fully verify the data, command, and clock paths within the chip will reduce the time required to perform validation on the extracted netlist because each block is self contained. The second advantage of this modular structure is that varying densities of memory chips can be easily constructed for varying applications. DRAM chips utilizing both proximity communication and this modular structure can simply be glued directly over their application with the correct density and I/O count. This has the possibility of revolutionizing the way chips access off-chip data. Instead of driving data requests away from the central circuitry of an integrated circuit to the memory channel, it is possible to simply send signals up towards the memory chip. This approach provides the exact memory requirement at the exact place it is required, reducing the access latency considerably.

## V. SUMMARY

Developing a wide I/O DRAM architecture that is suitable for proximity communication necessitates the communication channel to be moved to the side of the DRAM chip. This enables a proximity communication DRAM chip with 8 or 16 data pins. This modification requires limited design changes from current DRAM architectures.

A distributed page and bank structure was developed to enable the possibility of using proximity communication with 32 data pins. The architecture utilized the standard main memory page size specification of 8k, which allows the array power consumption to remain competitive with current and future DRAM architectures.

Reaching the use of 64 data pins required architectural changes that would not increase the manufacturing cost compared to current DRAM architectures. Three levels of metal above the memory capacitor is the projection for DRAM densities greater than 2 Gb. The wide I/O architecture allows the metal stack to remain at two levels of metal above the memory capacitor without increasing the chip size. The reduction of projected metal usage enables a significant cost advantage when compared to other DRAM architectures. A new column structure was introduced that will aide in the development of a proximity communication enabled DRAM architecture that utilizes $\geq 64$ data pins.

The wide I/O DRAM architecture utilizing proximity communication enables several technological advantages over existing DRAM architectures. Fixing the page size and increasing the I/O count through the wide I/O DRAM architecture allows for an energy efficient DRAM architecture. Fig. 5 shows the relative energy per bit estimates for DRAM chips utilizing proximity communication.

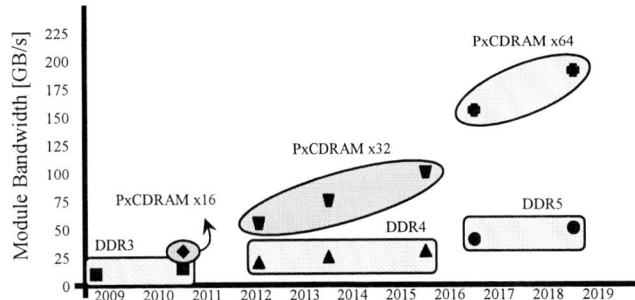

Figure 6. Module bandwidth comparison of current and future main memory compared to a main memory chips using proximity communication.

Current commodity DRAM chips have poor energy efficiency due to only using 64 data bits of the 8k bits accessed per page. The wide I/O architecture increases the number of bits accessed per page to 512, which significantly increases the energy efficiency of DRAM chips.

Although it is possible to only access one proximity communication DRAM chip to supply the full 64 bytes of data to the memory controller, it is also possible to increase the amount of data accessed by increasing the memory channel width. The projected bandwidth trend shown in Fig. 6 clearly shows the advantage of using proximity communication DRAM over current and future DRAM technologies.

## REFERENCES

[1] J. Hennessy, D. Patterson, Computer Architecture A Quantitative Approach, 4th ed., Morgan Kaufmann Publishers, San Francisco, 2007. ISBN 978-0-12-370490-0

[2] D. Rhosen, "The evolution of DDR," VIA Technology Forum, 2005.

[3] D. Patterson, "Latency lags bandwidth," Communications of the ACM, vol. 47, Issue 10, pp. 71-75, October 2004.

[4] D. Klein, "The future of memory and sorage: closing the gap," Microsoft WinHEC 2007, May 2007.

[5] Rambus, "Challenges and solutions for future main memory," http://www.rambus.com/assets/documents/products/future_main_memor y_whitepaper.pdf, May 2009.

[6] R. Drost, R. Hopkins, I. Sutherland, "Proximity communication," Proceedings of the IEEE 2003 Custom Integrated Circuits Conference, vol. 39, issue 9, pp. 469-472, September 2003.

[7] Q. Harvard, "Wide I/O DRAM architecture utilizing proximity communiation," Master's thesis, Boise State University, December 2009.

[8] D. Salzman, T. Knight, "Capacitively coupled multichip modules," Multichip Module Conference Proceedings, pp. 487-494, April 1994.

[9] K. Kilbuck, "Main memory technology direction," Microsoft WinHEC 2007, May 2007.

[10] R. Drost, R. Ho, R. Hopkins, I. Sutherland, "Electronic alignment for proximity communication," IEEE International Solid State Circuits Conference, vol. 1, pp. 144-145, February 2004.

[11] Samsung Semiconductor Inc. Various Datasheets: http://www.samsung.com/global/business/semiconductor/productList.do ?fmly_id=690

[12] International Technology Roadmap for Semiconductor, 2007 Edition, http://www.itrs.net/Links/2007ITRS/Home2007.htm, 2007.

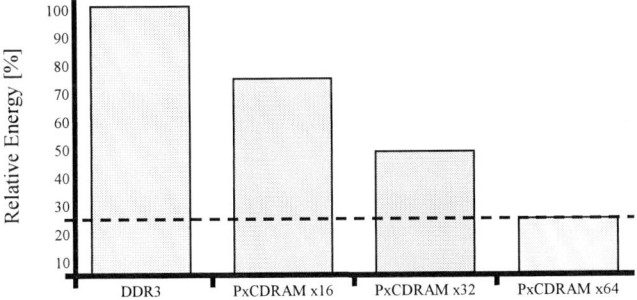

Figure 5. Energy per bit comparison.

978-1-4244-6572-9/10 $26.00 © 2010 IEEE

# A Low Noise Low Power DC Coupled Sensor Amplifier With Offset Cancellation

Hari Krishnan Krishnamurthy, Dirk Robinson, Dave M. Rector and George S. La Rue

Washington State University
Pullman, WA, USA 99164-2780
larue@eecs.wsu.edu

*Abstract*—A 16-channel sensor amplifier was designed in the Jazz 0.18μm CMOS process. The sensor amplifier has programmable gain from 20 to 2000 and a DC offset cancellation of ±0.3V using bulk voltage control. A charge pump, band gap reference and a resistor string DAC were designed for bulk voltage control. Input referred noise of 0.465 μV was achieved at a gain of 2000. The power required per channel was 900 μW and the supply voltages were ±1 V. Simulation results show the total harmonic distortion to be 0.017% and signal-to-noise-and-distortion (SINAD) ratio is 78 dB.

*Keywords- sensor amplifier, DC offset cancellation, charge pump, low noise, low power*

## I. INTRODUCTION

Biomedical research over the years has always had a need for multi-channel amplifiers to amplify the weak signals that are observed in different areas of research. This paper describes the development of a multi-channel integrated sensor IC with low noise, low power and DC cancellation to be used to detect and study the neural signals from the brain of small rodents. Other CMOS multi-channel integrated circuits have been reported for similar applications. These circuits either have very high power consumption or a high noise level [2] which will severely affect the accuracy of the signal being detected. High distortion [4] and large layout area [4] are also some disadvantages of comparable systems. Noise and power dissipation of some designs [3-4] are low but they lack programmable gain and use AC coupling which filters low frequency components.

## II. SENSOR AMPLIFIER ARCHITECTURE

The system architecture of the sensor amplifier is shown in Fig. 1. The 16 channels are multiplexed into an ADC and control signals are used to set the gain on the individual channels, calibrate the ADC and cycle the multiplexer through a set of channels [1]. The gain on each channel can be individually set from the control block.

An electrode is attached to the skull as a reference signal and can have up to ±0.3 volts DC offset compared to signals in the electrode array. Most of the multi-channel ICs available employ AC coupling to remove the DC offset. However this technique usually leads to the use of large off-chip capacitors to provide a low enough cut-off frequency and filters out data of

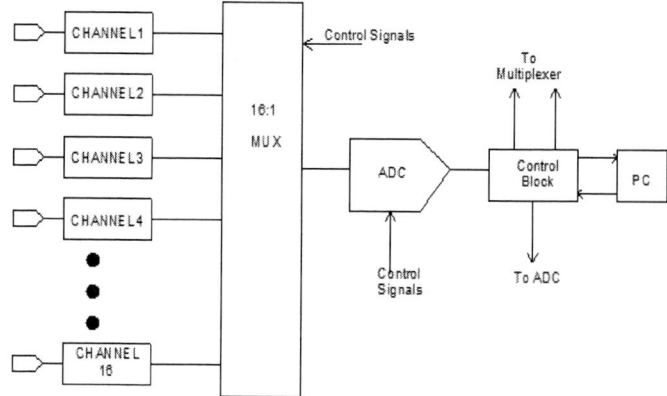

Figure 1. System architecture of sensor amplifier

interest in some EEG studies. This design adjusts the bulk of the input transistors to remove DC offset.

An electrode is attached to the skull as a reference signal and can have up to ±0.3 volts DC offset compared to signals in the electrode array. Most of the multi-channel ICs available employ AC coupling to remove the DC offset. However this technique usually leads to the use of large off-chip capacitors to provide a low enough cut-off frequency and filters out data of interest in some EEG studies. This design adjusts the bulk of the input transistors to remove DC offset.

Linearity is one of the important requirements of the sensor amplifier. The resistors, capacitors, switches and op-amp's all contribute to the non-linearity of the system. The op-amps were designed with very high open loop gain to provide high linearity. Also, poly resistors and MIM capacitors which have high linearity are used. The linearity of the channel is characterized by the total harmonic distortion (THD) discussed in the simulations section.

## III. CHANNEL ARCHITECTURE

The channel architecture is shown in Fig. 2. The programmable gain on each channel is split across two stages. The first stage has gains of 10, 25 and 50 while the second has gains of 2, 8 and 20. The programmable gain is achieved by connecting different resistors to the feedback loop using FET switches. A RC filter with a cut off frequency of 7 KHz followed the first two stages of amplification. The signal was then buffered before outputting to the multiplexer and ADC.

978-1-4244-6572-9/10 $26.00 © 2010 IEEE

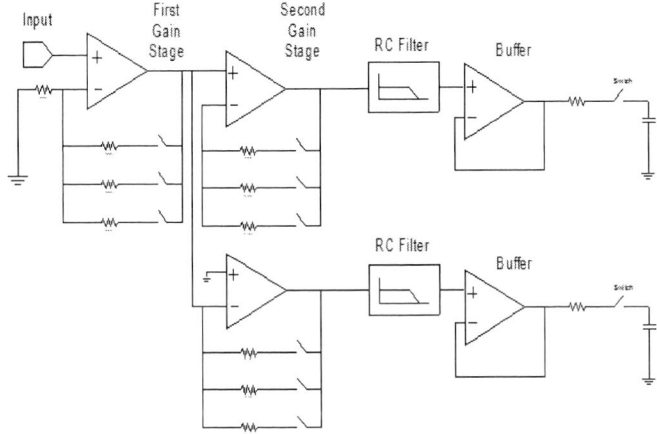

Figure 2.   Channel architecture

The first stage op amp is shown in Fig. 3. The design of the first stage op-amp with low noise was critical to overall noise performance. The input pair of the first stage op-amp is a set of PFET transistors with isolated bulks. DC offset cancellation was achieved by controlling the bulk voltages of the input pair of PFET transistors. In addition to being able to control their bulk voltages, PFETs have lower flicker noise than NFETs. Chopping was used in [1] to reduce flicker noise in the load and output transistors but adds complexity. Here resistors were added in series with the sources of the NFET load transistors to reduce their noise and provide better noise reduction with much lower complexity.

An input offset voltage occurs if the threshold voltages of the two transistors in the input differential pair of the first stage amplifier do not match. Forcing a mismatch of the threshold voltages can therefore cancel an offset voltage of an input signal. The threshold voltages of the PFET transistors are individually adjusted by controlling their bulk voltages. The threshold voltage mismatch needs to have a magnitude of 0.3V to achieve the desired offset cancellation of ±0.3V. Reverse biasing the source-bulk junction raises the threshold. Although it is possible to obtain a threshold mismatch of 0.3V with only reverse biasing, higher thresholds make operation with a 1V supply more difficult. Reverse biasing is used for offsets up to about 0.2V and then one of the PFET source-bulk junctions is forward biased to lower the threshold by up to ~0.1V. In order to raise the threshold voltage by 0.2V the bulk voltage needs to be 1.5V. This voltage is larger than the positive supply voltage and is generated on-chip with a charge pump.

The voltages of the bulks of the input pair were varied using a 5-bit decoder and resistor string DAC operating between +1.5V and ground. A switching matrix is used with the decoder to set the appropriate voltage at each of the bulk terminals. The DAC architecture used to set the bulk voltages is shown in Fig. 4. The DAC consists of course adjustment resistors as well as fine adjustment resistors. A single resistor string is used to obtain all of the voltages required. One decoder and set of switches applies the course DAC voltage to the bulk of one FET. Another decoder and switches outputs the fine voltages to the bulk of the other FET. Together the

resolution in offset voltage cancellation is sufficient at the highest gain to keep the output from saturating any of the amplifiers.

The second gain stage of the op amp is for amplification. Fig. 5 shows the op-amp used in the second stage. Since the gain of the first stage is at least 10, the noise contribution of the second stage is not critical and the second stage op-amp power can be low. The offset is not entirely canceled by the bulk voltages and results in a common mode voltage being applied to the second op-amp. The op-amps used in this stage have NFET's as input pairs in the second stage since a higher input common mode range is needed. A single band-gap reference is used to provide the voltage with enough current to drive the entire set of resistor string DACs for all 16 channels. A 1.5 V reference voltage is required.

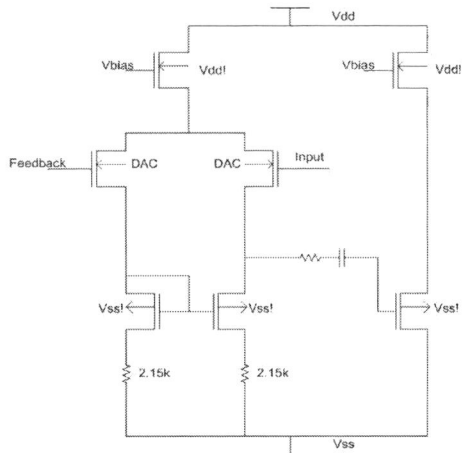

Figure 3.   First Stage Op-amp

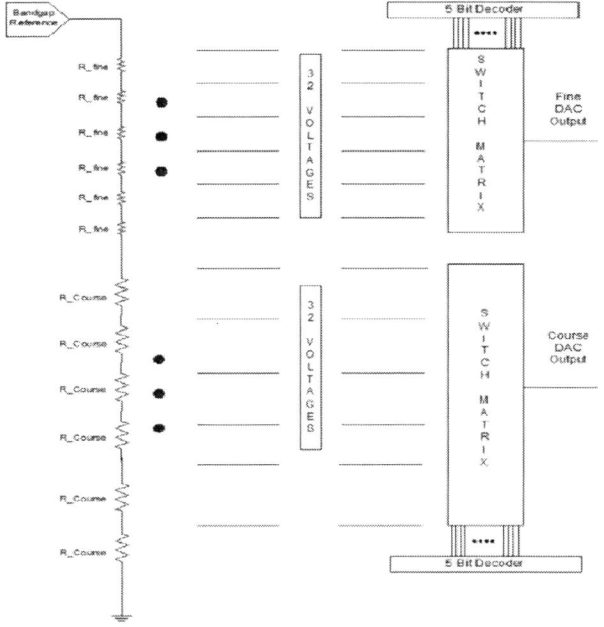

Figure 4.   DAC architecture

978-1-4244-6572-9/10 $26.00 © 2010 IEEE          45

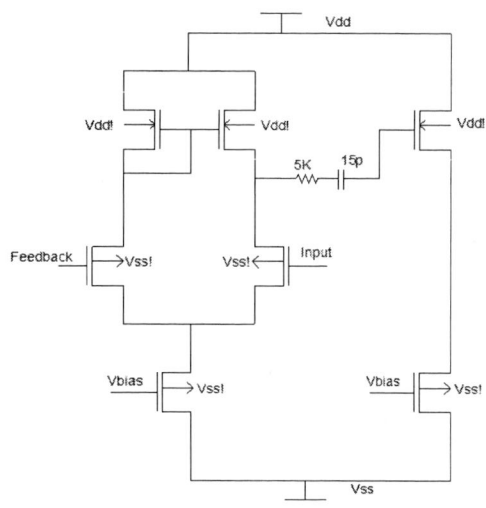

Figure 5. Second stage op amp

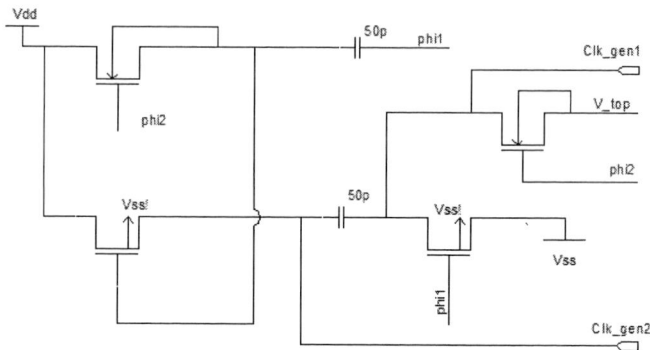

Figure 8. High voltage clock generator

## IV. CHARGE PUMP

The model of the charge pump design [5] is shown in Fig. 6. During the initial clock phase phi1 the capacitor C_ch is charged between $V_{dd}$ and $V_{ss}$. During the clock phase of phi 2 the bottom plate of C_ch is moved to $V_{ss}$ from $V_{dd}$ and hence the charge on the top plate increases to $2V_{dd}-V_{ss}$. The difference between the plates is then passed on to C_out. After several clock cycles the required voltage output across the capacitor C_out is obtained. The circuit realization is shown in Fig. 7. Special clocks are generated to provide the clocking signals to the transistors handling the higher voltages. The schematic of the high voltage clock generation circuit [5] is shown in Fig. 8.

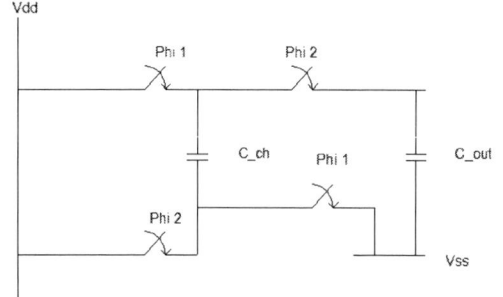

Figure 6. Charge pump block diagram

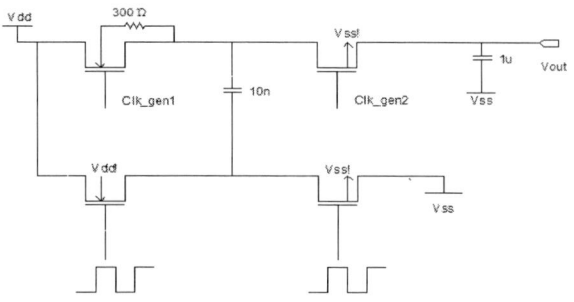

Figure 7. Charge pump schematic

## V. SIMULATIONS

The variation of the noise with the gain of the channel is shown in Table I. It is observed that the integrated input referred noise decreases as the gain of the channel is increased from 20 to 2000. The noise is the least at a gain of 2000 with no input DC offset applied and its value is 0.465 μV. The main noise contributors are from the first stage of the op-amp as it has a minimum gain of 10. The variation of the noise component with the DC offset at the input is shown in Table II. The gain is maintained at 2000 while varying the offsets in Table II. The integrated input referred noise increases with increasing magnitude of the input referred offset voltage.

The total harmonic distortion was found to be 0.017%. The signal-to-noise-and-distortion (SINAD) ratio is found by first calculating the RMS of the harmonics, $v_h$ with

$$v_h = \sqrt{v_2^2 + v_3^2 + v_4^2 + v_5^2} = 92.04\mu V \tag{1}$$

Then the SINAD is calculated as

$$SINAD = 20\log\sqrt{\frac{v_1^2}{v_h^2 + v_n^2}} = 78.7 dB \tag{2}$$

where $v_n$ is the RMS of the noise voltage.

TABLE I.    GAIN VS INPUT REFERRED NOISE

| Offset | Noise (in V) | Input referred noise (in μV) |
|--------|--------------|------------------------------|
| 0.1    | 0.00138      | 0.69                         |
| 0.2    | 0.00166      | 0.83                         |
| 0.3    | 0.00184      | 0.92                         |
| -0.1   | 0.00129      | 0.65                         |
| -0.2   | 0.00176      | 0.88                         |
| -0.3   | 0.00181      | 0.91                         |

978-1-4244-6572-9/10 $26.00 © 2010 IEEE

TABLE II.    GAIN VS INPUT REFERRED NOISE

| Gain | Noise (in μv) | Input Referred Noise (in μv) |
|------|---------------|------------------------------|
| 20   | 21.6          | 1.079                        |
| 160  | 102.6         | 0.6412                       |
| 800  | 399.3         | 0.4790                       |
| 2000 | 991.1         | 0.4655                       |

## VI.    CONCLUSION

A low-noise multi-channel sensor amplifier design is presented with programmable gain from 20 to 2000. DC offsets up to ±0.3V can be canceled by controlling body biases of the input transistors. The power dissipation is 900 μW per channel with power supply voltages of ±1V in the Jazz 0.18μm CMOS process. The sensor amplifier achieves a SINAD of 78dB. The integrated input referred noise is less than 0.5μV at high gains and is achieved without using chopping.

## ACKNOWLEDGMENT

This work was supported in part by National Institute of Health, R01-MH060263 and R01-MH071830.

## REFERENCES

[1] W. Zheng and G.S. La Rue, " Low-power low-noise DC-coupled sensor amplifier IC," *IEEE Workshop in Microelectronics and Electron Devices*, Boise, ID, April 2008.

[2] I. Obeid, J.C. Morizio, K.A. Moxon, M.A.L. Nicolelis, and P.D. Wolf, "Two multichannel integrated circuits for neural recording and signal processing," *IEEE Trans. Of Biomed. Engin.* vol. 50, pp. 225-258. Feb. 2003.

[3] M. Dagtekin, W. Liu and R. Bashirullah, "A multi channel chopper modulated neural recording system," *Proc. IEEE Engineering in Medicine and Biology Society, pp. 757-760,2001.*

[4] R. R. Harrison, C. Charles, "A Low-power low-noise CMOS amplifier for neural recording applications", *IEEE J. of Solid-State Circuits*, Vol. 38, no. 6, pp. 958 – 965, June, 2003.

[5] J. Silva-Martinez, "A switched capacitor double voltage generator" *Proc. IEEE Midwest Symp. on Circuits and Systems*, Lafayette, LA, pp. 177-180, Aug. 1994.

[6] B. Razavi, *Design of analog CMOS integrated circuits*, New York, NY; McGraw-Hill, 2000.

# Integration of a New Column-Parallel ADC Technology on CMOS Image Sensor

Fan Z. Nelson

Electrical and Computer Engineering
University of Idaho,
Moscow, Idaho, U.S.A
fan.nelson@vandals.uidaho.edu

Suat U. Ay

Electrical and Computer Engineering
University of Idaho,
Moscow, Idaho, U.S.A
suatay@uidaho.edu

*Abstract*—A new analog-to-digital converter (ADC) technology called Single-slope look ahead ramp (SSLAR) analog-to-digital converter (ADC) was proposed for column-parallel CMOS image sensors. Additionally, a corresponding programmable ramp generator for SSLAR ADC was also designed in such a way that it only allows flexible code hopping (between 0 and 127 least significant bit (LSB)), code fall back and look-ahead operations in column-parallel ramp ADC. This new ADC technology is able to provide conversion speed improvement depending on individual image information with less than 1.0% image quality degradation. Simulation demonstrated that the conversion speed of this new ADC technology is 4-5x higher than a traditional single-slope ADC with minimal circuitry for processing 8-bit standard gray images as well as 3-4x for standard CIF videos. For processing higher resolution images, the conversion speed may further increase. A prototype chip using the 8-bit SSLAR ADC architecture was realized in a 0.5μm, 2P3M, CMOS process with a layout area of 8.2mm$^2$.

*Keywords*- Analog to Digital Converter; ADC; single-slope look ahead ramp ADC; single-slope ramp ADC; integrating ADC.

## I. INTRODUCTION

Currently, several different types of column parallel ADCs (Analog to Digital Converters) have been used in CMOS (Complementary Metal Oxide Semiconductor) image sensors. A 12-bit column parallel ADC with accelerated ramp [1] was published in 2005 and exploits a way to reduce the number of ramp steps based on the nature of photon shot noise limitation. The results show that the image quality is not affected by the accelerated ramp modulation when shot noise ($M_{shot}$) is equal or less than 0 dB. Additionally, conversion speed can be greatly enhanced by reducing the number of ramp cycles without degradation of image quality.

The concept of "a new simultaneous multislope (SMS) analog-digital converter (ADC) architecture" [2] came out one year later. The proposed SMS ADC architecture works like a column parallel single-slope ADC except in that it has one comparison phase and one slope phase. Furthermore, several slopes are used in parallel during the slope phase, where each slope covers a part of the total signal swing. The result indicates that, when compared to the single-slope ramp (SSR) ADC, SMS ADC has a conversion speedup approximately 2 times and requires only a small amount of extra circuitry. The disadvantage is a relative low conversion speedup improvement.

An additional method of increasing conversion speed has been suggested by using a new column-parallel ADC architecture based on a multiple-ramp single-slope (MRSS) ADC [3, 4] which improves upon the techniques in [2]. Further increases in conversion speed can be made by combining the MRSS ADC architecture with the concept of exploiting the amplitude-dependent characteristic of photo shot noise present in image signals [1]. But the disadvantage here is that it is difficult to implement such a large number of ramps. First, each output of the ramp generator must be buffered in order to drive the capacitive load presented by a larger number of comparators. Since the number of comparators connected to each ramp is signal-dependent, each buffer has to be dimensioned for worst case situations, where it has to drive all of the comparators. Therefore, there is tradeoff between the increased speed obtained by using a large number of ramps and increased power dissipation of using many ramp buffers.

Two-step single-slope ADCs based on multiple ramp signals have been reported in [2] and [4]. However, these two-step single-slope ADCs require several ramp generators corresponding to the number of coarse ADC steps. This architecture has increased area and power consumption due to the multiple ramp generators and multiple ramp signal lines. Therefore, coarse ADC resolution is limited to only 2 or 3 with little speed improvement. A new two-step SS ADC using a single ramp generator for the coarse and fine ADCs has been proposed [5, 6]. This technique is free from ramp slope mismatch and high power consumption that were the bottlenecks of the conventional two-step SS ADCs with multiple ramp generators. The results demonstrate that the proposed ADC has ten times higher conversion speed without any significant increase in power consumption when compared to the conventional SS ADC.

In contrast with the techniques mentioned above, our method uses an alternative approach which also combines the statistical analysis of the images. In this paper, we introduce a new ADC algorithm and column parallel ADC architecture integrated on a column-parallel CMOS image sensor to reduce

978-1-4244-6572-9/10 $26.00 © 2010 IEEE

SSR-ADC's latency without degrading the image quality and integrity. The new ADC introduces code hopping, fall back, and code look-ahead operations considering the statistics on the sampled row signals. We call this architecture a single slope look-ahead (SSLAR) ADC architecture. This paper focuses on the introduction of the SSLAR ADC algorithm and its implementation and simulation using MATLAB as well as AMIS 0.5μm process to verify the proper operations.

This paper is organized as follows. Section II introduces the topology of the SSLAR ADC. A corresponding programmable ramp generator topology is introduced in section III. Design and simulation results are reported in section IV. Section V is dedicated for conclusion.

## II. SINGLE-SLOPE LOOK-AHEAD RAMP (SSLAR) ADC

In order to address the speed issue for single slope ramp (SSR) ADC, a new ADC algorithm and topology called single-slope look-ahead ramp (SSLAR-ADC) ADC was reported [7]. Correspondingly, a programmable ramp generator with changeable code hopping was also introduced. The new ADC algorithm was successfully implemented in MATLAB. Simulation on both standard gray images and CIF videos were conducted. It was determined that, by setting proper threshold and jump step, the algorithm results in increased ADC speed; improving frame rate of the CMOS image sensor without greatly degrading image quality [7]. For a single input, conversion can be greatly improved by using SSLAR ADC when a proper step size is chosen. For example, when a step size of 24 is chosen a speedup of approximately 6x can be achieved when compared to SSR ADC (refer to Fig.1-2). An additional 8-bit standard sample image "LENA" can also be observed in Fig.3.

Different jump steps were considered during simulation to plot both mean-square error (MSE) rated in percent of full scale and conversion speed up. It was determined that, for proper thresholds, jump step sizes between 4 and 6 LSB achieve an ADC speedup approximately 4-5x times and results in less than 1% image quality degradation for 8-bit standard gray image. In this case, jump step sizes between 4 and 6 can be claimed as an ideal jump step size range if 256×256 (8-bit) standard gray images need to be processed. For processing higher resolution images, the thresholds may also be increased little. In the same fashion, for proper thresholds, larger jump step sizes could be set for processing higher resolution images. In conclusion, changeable step sizes are necessary depending on individual image information. Thus, a fully-flexible programmable ramp generator with flexible code hopping is needed in order to fit this new SSLAR ADC algorithm well.

## III. PROGRAMMABLE RAMP GENERATOR DESIGN

In this section, only the digital circuitry will be described. The new programmable step SSLAR ADC composed of three major units is shown in Fig.4. They are the event detector (ED), controller unit (CONT) and the ramp-count generator (RCG) blocks. Step word N[6:0] can be programmed between 0 to 127 LSB for different applications.

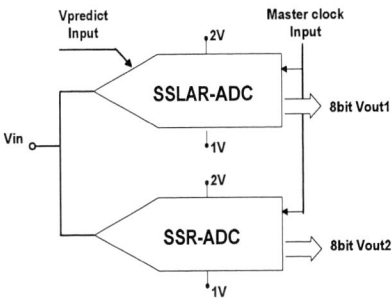

Figure 1. SSLAR-ADC Versus SSR-ADC with only one input.

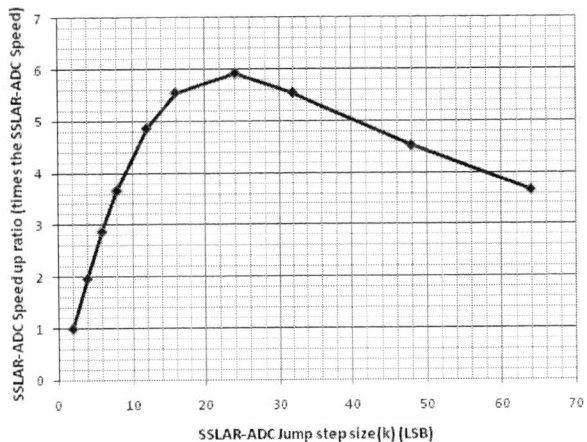

Figure 2. Speed up of the SSLAR-ADC compared to SSR-ADC with a single signal voltage input.

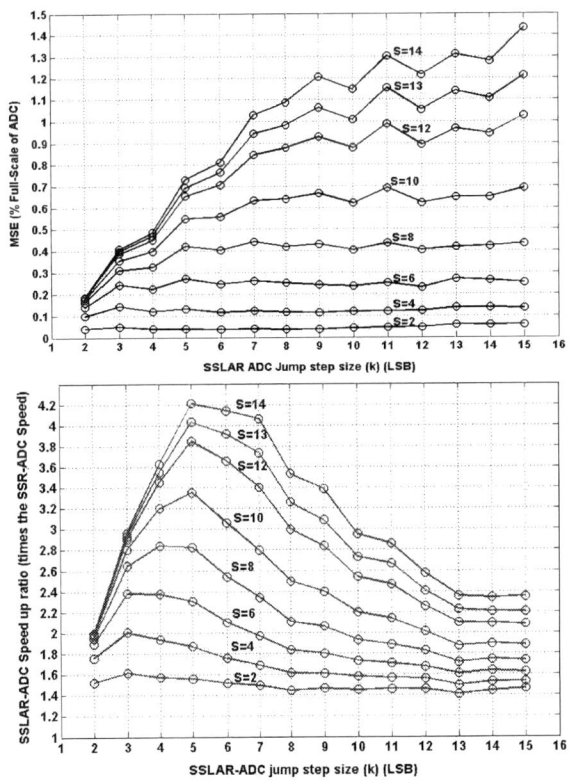

Figure 3. MSE versus jump steps for 8-bit "LENA" (upper figure) and ADC speed-up versus jump steps for 8-bit "LENA" (bottom figure).

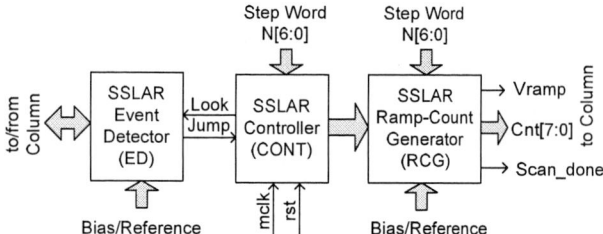

Figure 4. Programmable SSLAR ADC blocks.

## A. Event Detector (ED)

The event detector (ED) is connected to the column predictor circuit and generates the "jump" signal depending on the "look" signal and column predictor conditions. ED also receives analog reference and bias signals.

## B. The Controller Unit (CONT)

Controller unit (CONT) is central to the SSLAR ADC generates the "look" signal for the event detector (ED) and other control signal for ramp-count generator (RCG) blocks. CONT coordinates and implements the SSLAR ADC algorithm

A finite state machine (FSM) was used for generating six different operation states (refer to Fig.5 and Table.1) as well as internal and external control signals for CONT. Four control signals "rst" (external operation reset), "jump" (external jump signal from ED unit), "done" (internal counter done signal) and "mclk" (external master clock signal) determine the FSM states. FSM changes its state at rising edge of the master clock signal conditionally or unconditionally. At the same time, these six operation states also generate six additional control signals. Three of which are most important since they are externally fed to ED and RCG units. They are "look", "Lclk" and "C[2:0]" signals.

## C. SSLAR Ramp-Count Generator (RCG) Design

The ramp-count generator (RCG) unit generates the analog ramp signal and the associated 8-bit digital counter output. Its block diagram is shown in Fig.6. RCG is composed of two multiplexers, one carry look-ahead (CLA) full-adder with latch (FAL) block, two carry look-ahead subtractors and one 8-bit binary weighted charge scaling ramp generator block. Thus, look ahead, jump and fall back operations are controlled through the proper timing of the blocks.

Table I. Function description of six states for CONT unit.

| State | Function Description |
|-------|---------------------|
| S0 | Count starts at zero and ramp voltage starts at original point (original state). |
| S1 | Ramp voltage look-ahead k step size and counter look ahead k/2 step size from original. |
| S2 | Ramp voltage maintains the same, but counter jump rest k/2 step size. |
| S3 | Ramp voltage continues to look ahead k step size and counter look-ahead k/2 step size from previous start point. |
| S4 | Ramp voltage falls back k step size and counter falls back k/2 step size from previous start point due to unsuccessful look ahead. |
| S5 | Ramp voltage and counter raise one step size per time. |

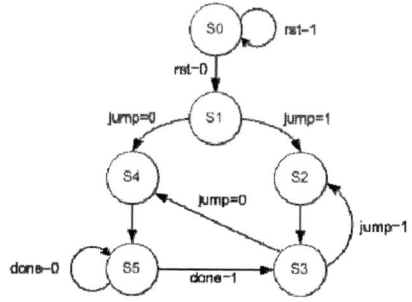

Figure 5. State machine diagram of the controller unit

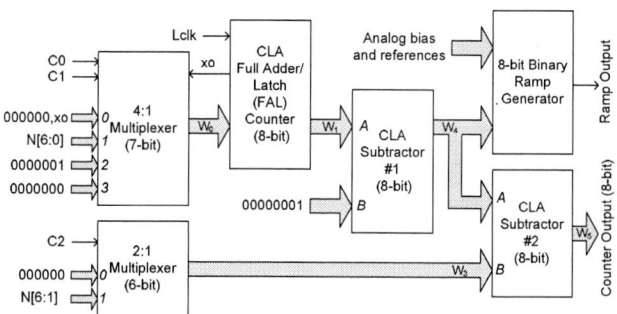

Figure 6. SSLAR ADC ramp-count generator unit block diagram

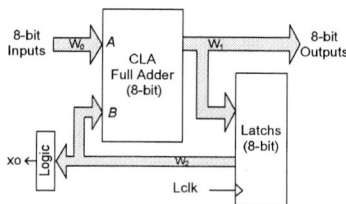

Figure 7. Block diagram of CLA full-adder and latch (FAL) unit

Block diagram CLA full-adder and latch (CLA-FAL) unit is the central part of the look-ahead and jump operation. It is shown in Fig.7. Latch clock (Lclk) is generated from CONT unit and fed to the RCG unit. One 8-bit input (A[7:0]) to CLA full-adder are provided by the 4:1 multiplexer unit and another 8-bit input (B[7:0]) are provided by the 8-bit latch outputs. CLA full-adder adds these two inputs to generate the CLA-FAL block output and feedback to the 8-bit latch. When Lclk signal is asserted from low to high, latch block holds the outputs of the CLA-FAL unit.

Multiplexer selection code ($C_0$, $C_1$, $C_2$) are generated by the CONT unit based on the state of FSM. For 4-to-1 multiplexer, when $C_0$= '0' and $C_1$= '0', xo value is passed; when $C_0$= '0' and $C_1$= '1', digital one is passed; when $C_0$= '1' and $C_1$= '0', the 7-bit step programming word (N[6:0]) is passed; input $C_0$= '1' and $C_1$= '1' was never selected and was set to zero. By taking upper 6 bit of the original step program word from 4-to-1 multiplexer, half step code word is attained from 2-to-1 multiplexer. 2-to-1 multiplexer always passes the half step programming words. When $C_2$= '1', it look ahead half step size; when $C_2$= '0', it jump rest half step size and implements the full step size counter increment operation. An 8-bit binary weighted capacitive digital to analog converter was used as ramp generator.

## IV. LAYOUT DESIGN AND SIMULATION RESULTS

Proposed design was fabricated in a 0.5μm 2P3M CMOS process. Supply voltage is 3.3V and layout of 8-bit ramp generator is shown in Fig.8. Core area SSLAR-ADC is approximately 0.54 mm².

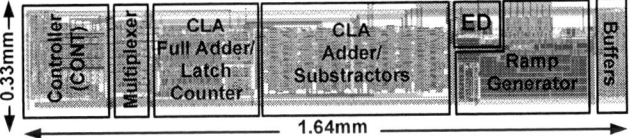

Figure 8. Layout of 8-bit programmable ramp generator.

Jump step size can be easily programmed between 0 and 127 LSB. Here, sample step size 8 was chosen and corresponding simulation results are shown in Fig.9. The jump or fall back operation consumes two clock cycles.

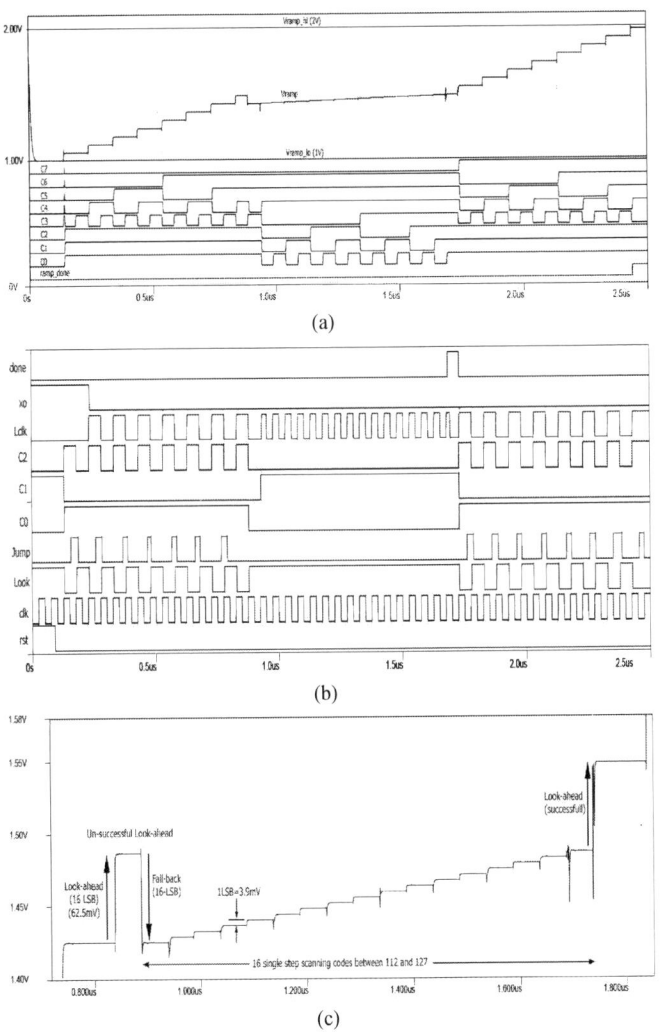

Figure 9. 8-bit full block simulation results of the SSLAR ADC's controller and ramp-count generator unit with step=16LSB, and jump signal is forced; (a) analog signals, (b) digital control signals (c) Zoomed ramp output voltage (Vramp) for failed look-ahead operation between codes 112 and 127.

## V. CONCLUSION

Compared to the single slope ramp (SSR) ADC, the single-slope look-ahead ramp (SSLAR-ADC) ADC's conversion speed was greatly enhanced due to skipping of code ranges where few pixels in that range lie without image degradation. A 12-bit ADC architecture could be easily expanded based on the 8-bit architecture and would be expected to enhance conversion speed even more. A potential downside to SSLAR architecture is the increased power consumption with extra circuitry. Table II contains a comparison of other available methods to our proposed technique. The proposed programmable ramp generator can attain variable step sizes which can optimize the image process quality as well as conversion speed in a user specific manner. Future directions could be a combination of photon shot noise limitation techniques with our SSLAR-ADC design and may enhance our ADC's conversion speed even more.

## VI. ACKNOWLEDGEMENT

We would like to thank Micron Technology Foundation for support and MOSIS fabrication service for allowing proposed design to be fabricated in a subsidized 0.5μm 2P3M CMOS run (PID#82059).

Table II. Available methods

| Ref. | Speedup (times) | Die Area (mm²) | Year | Process (μm) | Power |
|---|---|---|---|---|---|
| [2] | ~2 | extra circuitry | 2006 | 0.35 | N/A |
| [4] | MRSS:~3.3 MRMS:~4.4 | 5×5 | 2007 | 0.25 | 52mW |
| [6] | ~10 | 3.6×3.2 | 2009 | 0.35 | 36mW |
| [7] | BW (8b): ~5 | 2.0×4.1 | 2009 | 0.5 | 56mW |

## VII. REFERENCES

[1] Toshinori Otaka et al., "12-bit Column-parallel ADC with accelerated ramp,"Institute of electronnics, Information and communication engineers, vol.105. no. 185, pp. 35-38, 2005.

[2] Leif Lindgren, "A new simulataneous Multislop ADC architecture for array implemetation," *IEEE Trans. On Circuit Syst.II, Exp. Briefts*, vol.53, no.9, pp.921-925, Sep,2006.

[3] Snoeij, et al., "A CMOS Image sensor with a column-level multiple-ramp single-slope ADC," *ISSCC 2007*, pp 506,618,11-15 Feb.2007.

[4] M. F. Snoeij, et al., "Multiple-ramp column-parallel ADC architectures for CMOS image sensors," *IEEE J. Solid-State Circuit*, vol.42, no.12, pp.2998-3006, Dec.2007.

[5] J. Lee,et al., "A 10b column-wise two-step single-slope ADC for high-speed CMOS image sensor," in Proc. *IEEE Int. Image sensor Workshop, Ogunquit, ME*, pp. 196-199, Jun. 2007.

[6] S.Lim,et al., " A high-speed CMOS Image sensor with column-parallel two-step single-slope ADCs" *IEEE Trans. Electron Devices*,vol.56,no.3, pp.393-398, Mar.2009.

[7] F. Z. Nelson, M. N. Alam, and S. U. Ay, "A Single-slope look-ahead ramp (SSLAR) ADC for column parallel CMOS Image sensors," *IEEE Workshop on Microelectronics and Electron Devices, WMED 2009*, 3 April 2008 .

# Gain Error Correction for CMOS Image Sensor using Delta-Sigma Modulation

Kuangming Yap and R. Jacob Baker

Department of Electrical and Computer Engineering
Boise State University
Boise, ID, U.S.A.

**Abstract** – A delta-sigma modulation analog-to-digital converter (ADC) has many benefits over the use of a pipeline ADC in a CMOS image sensor. This includes lower power, noise reduction, ease of maximizing the input range, and simpler signal routing for large arrays. Multiple delta-sigma modulation ADC is required in a CMOS image sensor, one for each pixel column. Any voltage threshold mismatch between ADCs will introduce gain and offset error in its transfer function, which will lead to fix pattern noise. Correcting these gain and offset error for every ADCs in the image sensor will require a complex digital signal processor. Therefore, a technique to minimize the effects of gain error in a delta-sigma modulation ADC for CMOS image sensor is discussed.

*Keywords – Delta-Sigma Modulator, DSM, clocked comparator, 4 phase non-overlapping clock, CMOS Image Sensor ADC .*

## I. INTRODUCTION

The pipeline analog-to-digital converter (ADC) is usually used in today's CMOS image sensors to convert the analog pixel integration voltage to a digital code [1]. There may be one pipeline ADC in a CMOS image sensor that converts the color of each color pixel in the array: green, blue and red. As the pixel density in the CMOS image sensor increases, or the level of color quality increases, the pipeline ADC will require a larger layout area and consume more power. Furthermore, as the size and density of the CMOS image sensor increases, coupled noise and voltage variation from the long distance required to route the signal between pixel and the pipeline ADC becomes an issue.

A per column delta-sigma modulation (DSM) ADC sensing circuit was introduced as a potential solution to the problem [1]. A per column DSM ADC is susceptible to fixed pattern noise due to voltage threshold mismatch between adjacent DSM ADCs in the array. Voltage threshold mismatch causes offset and gain error in its transfer function.

Path switching was introduced as a method to reduce the effects of offset error in a per column DSM ADC [1], [2]. This paper will introduce a technique to help reduce the effects of gain error in a per column DSM ADC.

## II. DELTA-SIGMA MODULATION ADC SENSING OPERATION IN A CMOS IMAGE SENSOR

Fig. 1 illustrates the sensing scheme used by a DSM in a CMOS image sensor. There is one DSM for each column of pixels and the sensing occurs in parallel for each column.

The sensing scheme begins with sampling the pixel's reset signal onto the reset sample and hold capacitor, $C_{HR}$. This is accomplished by turning on the rowline and ResetN signals on one of the multiple rows of pixels in the array. Each pixel on the activated row will output its reset signal onto its respective column line. This reset signal is then sampled onto the $C_{HR}$ capacitor through the activated SHR switch. Once the reset signal is sampled, the rowline, ResetN, and SHR switch is turned off.

After the reset signal is sampled, the pixel is then exposed to light for a length of time. Once the exposure time has expired, the same rowline and SHI switch is turned on. This samples the image signal onto the image sample and hold capacitor, $C_{HI}$. The rowline and SHI switch is turned off once the image signal is sampled onto $C_{HI}$.

Next, the DSM will take the reset and image input signals that were sampled on the two capacitors and convert them to a digital code equivalent. The n-bit wide counter that is connected at the end of the DSM converts the digital output of the DSM into an n-bit wide digital code.

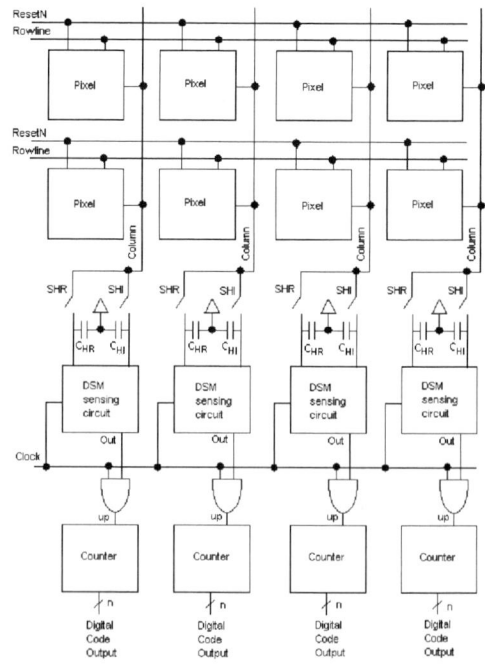

Figure 1. A typical 2x4 pixel diagram of a CMOS image sensor using DSM ADC [3].

978-1-4244-6572-9/10 $26.00 © 2010 IEEE

## III. A DELTA-SIGMA MODULATION ADC WITHOUT OFFSET AND GAIN ERROR CORRECTION

The basic DSM sensing circuit without any offset and gain error correction is seen in Fig. 2. It measures the difference between the two analog input signals, $V_{RESET}$ and $V_{IMAGE}$ with respect to a single reference input signal, $V_{REF}$. The DSM is clocked for N times over the entire sensing period. A simple two phase non-overlapping clock is required, where PHI1B is the inverse of the first phase and PHI2B is the inverse of the second phase. Its frequency is equal to the rate of the master clock denoted as $f_{PHI}$.

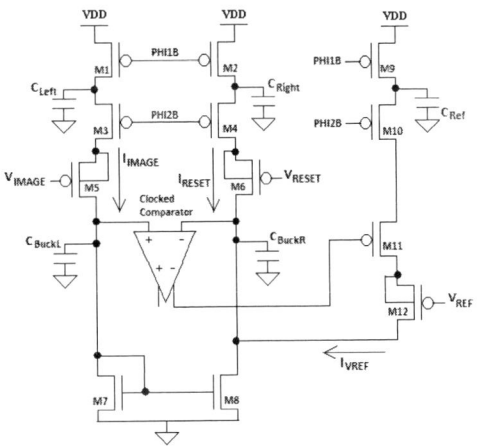

Figure 2. DSM ADC without gain and offset error correction circuitry.

At the end of the first phase of the clock, the clock comparator in the DSM measures the voltages on capacitor $C_{BUCKL}$ and $C_{BUCKR}$ and turns on M11 for the subsequent phase if the voltage on $C_{BUCKR}$ is lower than the voltage on $C_{BUCKL}$. This occurs for M times over the entire sensing period.

During the first phase of the clock, $C_{LEFT}$, $C_{RIGHT}$, and $C_{REFf}$ are set to VDD. On the next phase of the clock, the image current, $I_{IMAGE}$ and reset current, $I_{RESET}$ flows into $C_{BUCKL}$ and $C_{BUCKR}$ respectively.

$$I_{IMAGE} = C_{LEFT}f_{PHI}(VDD - V_{IMAGE} - V_{th,M5}) \quad (1)$$

$$I_{RESET} = C_{RIGHT}f_{PHI}(VDD - V_{RESET} - V_{th,M6}) \quad (2)$$

$V_{th,M5}$ and $V_{th,M6}$ is the threshold voltage of M5 and M6. If M11 is turned on, a reference current, $I_{VREF}$ will flow into $C_{BUCKR}$ and its magnitude average over the whole sensing period is

$$I_{VREF1} = \frac{M}{N}C_{REF}f_{PHI}(VDD - V_{REF} - V_{th,M12}) \quad (3)$$

$V_{th,M12}$ is the threshold voltage of M12.

The digital code representation of the analog input signals with respect to the reference signal can be found by summing the current into the $C_{BUCKR}$ capacitor.

$$M = N\left(\frac{C_{LEFT}(VDD - V_{IMAGE} - V_{th,M5}) - C_{RIGHT}(VDD - V_{RESET} - V_{th,M6})}{C_{REF}(VDD - V_{REF} - V_{th,M12})}\right) \quad (4)$$

The input-output transfer function of this DSM is prone offset and gain error if voltage threshold mismatches exist between the many DSM ADCs in the array. If a voltage threshold mismatch on M5 and M6 exist, an offset error will

occur. On the other hand if a voltage threshold mismatch on M12 exist, a gain error will occur. Therefore, CMOS image sensor that uses this DSM will be susceptible to fixed pattern noise because the transfer functions between the many DSM ADCs in the array will be different.

## IV. DELTA-SIGMA MODULATION ADC WITH GAIN ERROR CORRECTION

As mentioned earlier, a per column DSM ADC with path switching was introduced to reduce the effects of offset error [1]. A DSM sensing circuit with gain error correction is shown in Fig. 3. This DSM measures the difference between the two analog input signals, $V_{RESET}$ and $V_{SIGNAL}$ with respect to two reference input signals, $V_{REF1}$ and $V_{REF2}$ instead. The DSM is also clocked for N times over the entire sensing period.

A 4-phase non-overlapping clock signal is required for this design, Fig. 4. The PHI1, PHI2, PHI3, and PHI4 signals are the four non-overlapping clock phases and its complement signals are PHI1B, PHI2B, PHI3B, and PHI4B. Its frequency is 1/4th the rate of the master clock denoted as $f_{PHI}$.

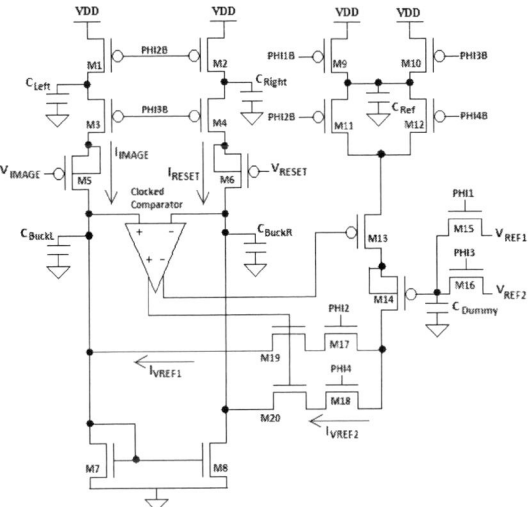

Figure 3. DSM ADC with gain error correction circuitry.

The clock comparator in this DSM also measures the voltages on capacitor $C_{BUCKL}$ and $C_{BUCKR}$ at the end of the first phase of the clock but it turns on M13, M19 and M20 for the remaining three clock phases instead if the voltage on $C_{BUCKR}$ is lower than the voltage on $C_{BUCKL}$. The clock comparator turns on M13, M19 and M20 for M times over the entire sensing period.

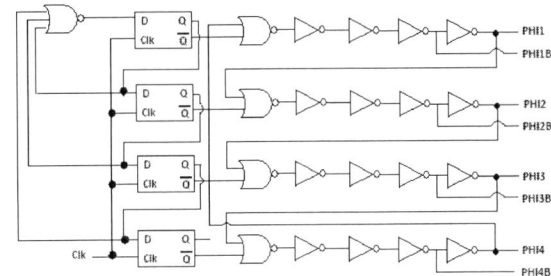

Figure 4. A schematic diagram that shows the 4 non-overlapping clock phases.

A dummy capacitor $C_{DUMMY}$ is added to the gate of M14 as a mean to reduce the effects of charge injection and clock feedthrough when PHI1 and PHI3 signals transitions from high to low. The reference signal voltages needs to be on the gate of M14 a phase earlier and stays unchanged for the whole duration of the subsequent phase. This is to prevent any error in the magnitude of the two reference current, $I_{VREF1}$ and $I_{VREF2}$.

During the first phase of the clock signal, M9 turns on and sets the voltage on $C_{REF}$ to VDD. At the same time, M15 turns on and allows the reference signal $V_{REF1}$ to propagate to the gate of M14.

On the next phase of the clock, M11 turns on and the first reference current, $I_{VREF1}$ will flow from $C_{REF}$ to $C_{BUCKL}$ if M13, M17 and M18 is turned on by the comparator. The average reference current that flows into $C_{BUCKL}$ is

$$I_{VREF1} = \frac{M}{N} C_{REF} \frac{f_{PHI}}{4} \left( VDD - V_{REF1} - V_{th,M14} \right) \quad (5)$$

$V_{th,M14}$ is the threshold voltage of M14. On the same clock phase, the voltage on $C_{LEFT}$ and $C_{RIGHT}$ are set to VDD.

On the third phase of the clock, $C_{REF}$ is force to VDD and the second reference voltage signal, $V_{REF2}$ is propagated to the gate M14. During this phase, the image current, $I_{IMAGE}$ and reset current, $I_{RESET}$ will flow to $C_{BUCKL}$ and $C_{BUCKR}$ respectively.

$$I_{IMAGE} = C_{LEFT} \frac{f_{PHI}}{4} \left( VDD - V_{IMAGE} - V_{th,M5} \right) \quad (6)$$

$$I_{RESET} = C_{RIGHT} \frac{f_{PHI}}{4} \left( VDD - V_{RESET} - V_{th,M6} \right) \quad (7)$$

On the last phase of the clock, the second reference current, $I_{VREF2}$ will flow from $C_{REF}$ to $C_{BUCKR}$ if M13, M17 and M18 remains on. The average reference current that flows into $C_{BUCKR}$ is

$$I_{VREF2} = \frac{M}{N} C_{REF} \frac{f_{PHI}}{4} \left( VDD - V_{REF2} - V_{th,M14} \right) \quad (8)$$

Over N clocked cycles, the sum of current into $C_{BUCKR}$ is ideally zero and the digital relationship between the analog input signals with respect to the reference signals is

$$M = N \left( \frac{C_{LEFT}(VDD - V_{IMAGE} - V_{th,M5}) - C_{RIGHT}(VDD - V_{RESET} - V_{th,M6})}{C_{REF}(V_{REF1} - V_{REF2})} \right) \quad (9)$$

The transfer function for this DSM shows that it is robust to any gain error cause by voltage threshold mismatch. This is because the denominator of the transfer function does not contain the voltage threshold of any transistor in the DSM. The gain of the DSM is controlled by the two reference voltage signals, $V_{REF1}$ and $V_{REF2}$. The difference between $V_{REF1}$ and $V_{REF2}$ determines the gain of the DSM.

However, this DSM is prone to offset error as the nominator of the transfer function contains the voltage threshold of M5 and M6. Path switching methodology can be applied to this DSM to reduce the effects of offset error [1]. Fig. 5 illustrated the DSM with both offset and gain error correction.

The sensing period of the DSM is now divided into 2 equal halves. Each halves is N/2 clock cycles long. On the first half of the sensing period, control signals SLT and SLB is set to

VDD and VSS respectively. During this sensing period, the digital code representation of the analog input signals with respect to the two reference input signals, $M_{t=1}$ is

$$M_{t=1} = \frac{N}{2} \left( \frac{C_{LEFT}(VDD - V_{IMAGE} - V_{th,M5}) - C_{RIGHT}(VDD - V_{RESET} - V_{th,M6})}{C_{REF}(V_{REF1} - V_{REF2})} \right) \quad (10)$$

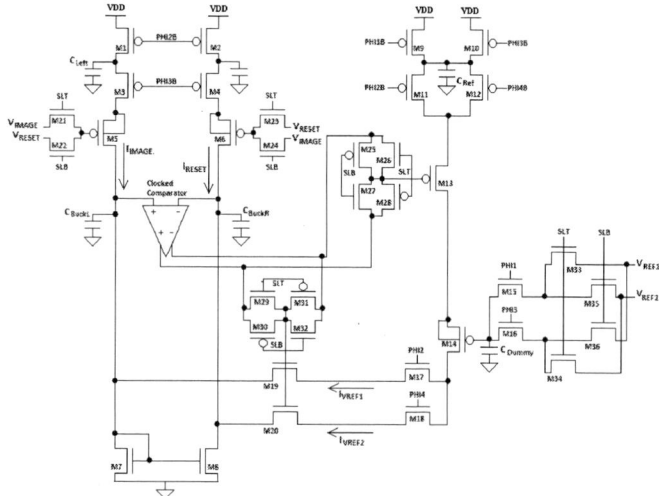

Figure 5. DSM ADC with gain and offset error correction.

On the next half of the sensing period, the controls signals SLT and SLB is set to VSS and VDD respectively. The digital code representation of the analog input signals with respect to the two reference input signals for the second half of the sensing period, $M_{t=2}$ is

$$M_{t=2} = \frac{N}{2} \left( \frac{C_{RIGHT}(VDD - V_{IMAGE} - V_{th,M6}) - C_{LEFT}(VDD - V_{RESET} - V_{th,M5})}{C_{REF}(V_{REF1} - V_{REF2})} \right) \quad (11)$$

At the end of the sensing period, the digital output code for the two halves of the sensing period is added together and the final digital output code for the DSM with both gain and offset correction is

$$M_{t=1} + M_{t=2} = \frac{N}{2} \left( \frac{(C_{LEFT} + C_{RIGHT})(V_{RESET} - V_{IMAGE})}{C_{REF}(V_{REF1} - V_{REF2})} \right) \quad (12)$$

The input-output transfer function of this DSM does not contain the voltage threshold of any transistor in the DSM. This means that the DSM is robust to any offset or gain error cause by voltage threshold mismatches. However, a gain error might still occur if the capacitor ratio between the sum of $C_{LEFT}$ and $C_{RIGHT}$ and $C_{REF}$ is not identical between the many DSMs in the array.

The least significant bit voltage, $V_{LSB}$ for the DSM with both offset and gain error correction can be approximated by

$$V_{LSB} = \frac{1}{N} \left( \frac{C_{REF}}{(C_{LEFT} + C_{RIGHT})} \right) \left( \frac{2}{(V_{REF1} - V_{REF2})} \right) \quad (13)$$

The bit accuracy increases linearly with the number of clock cycles, N during the sensing period.

## V. SIMULATION

A voltage source was added in series with the gate of a transistor to simulate a variation in its threshold voltage. A simulation is ran on the DSM without offset and gain error correction like in Fig. 2. A voltage threshold offset on M5

978-1-4244-6572-9/10 $26.00 © 2010 IEEE

causes the transfer to either shift upwards or downwards from ideal and it is known as offset error. A negative offset on M5 causes the curve to shift upwards and vice-versa. On the other hand, a voltage threshold offset on M12 causes the slope of the transfer function to either increase or decrease from ideal and this is gain error. A negative offset on M12 decreases the slope.

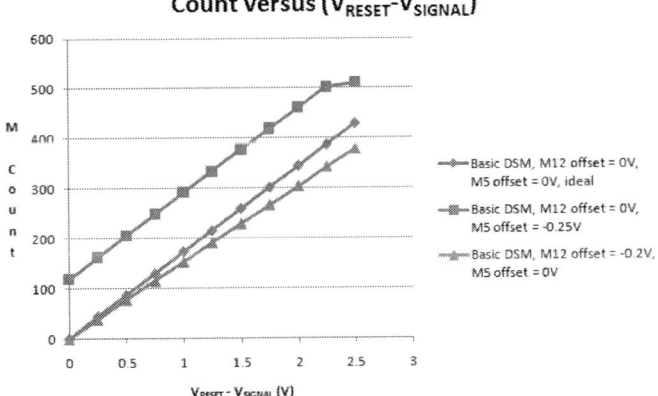

Figure 6. Simulation results for the DSM without offset and gain error correction. Different voltage offsets are applied to M5 and M12.

A simulation was run on the DSM with only gain error correction. Voltage offsets is placed in series with the gate of M14 to simulate voltage threshold variation that leads to gain error in its transfer function. Fig. 7 shows that the slope of the transfer function for this DSM does not change with different offsets on M13.

A 0.25V voltage offset was added in series with the gate of M5 and its transfer function is shifted downwards from ideal. As expected, the DSM with only gain error correction is susceptible to offset error.

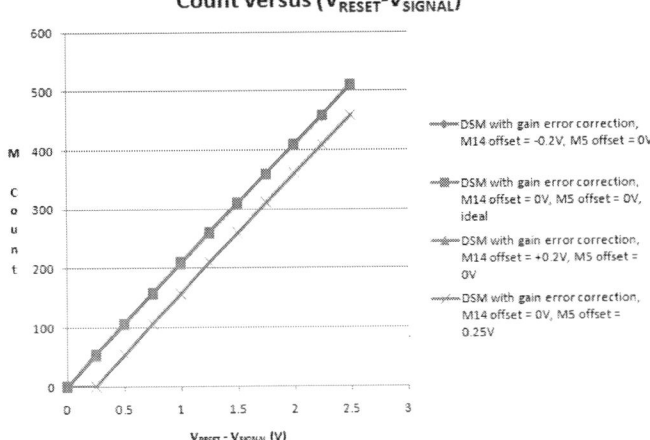

Figure 7. Simulation results for the DSM with only gain error correction. Different voltage offsets are applied to M5 and M14.

Fig. 8 shows that the DSM with both offset and gain error correction is robust to offset and gain error. Voltage offsets was added in series with the gate M14 and M5 to simulate gain and offset error. The transfer function with both voltage offsets looks identical to the transfer function of an ideal DSM except for the front and tail end. The magnitude of the deviation is equal to the amount of voltage offset applied to the gate of M5. This means the DSM input dynamic range is reduced by 0.5V or twice the amount of voltage threshold mismatch on M5.

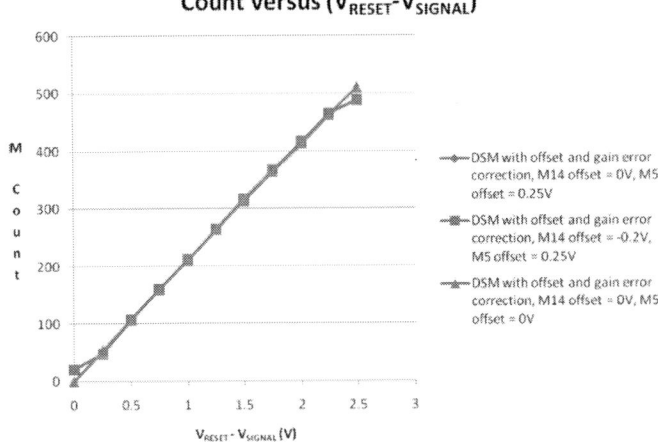

Figure 8. Simulation results for the DSM with both offset gain error correction. Different voltage offsets are applied to M5 and M14.

## VI.  SUMMARY AND CONCLUSIONS

The DSM proposed is robust to any voltage threshold mismatch that causes gain error. Path switching technique can be applied to this DSM which will cancel both offset and gain error. This will further reduce fixed pattern noise in a CMOS image sensor using per column DSM ADC because the transfer function of every DSM ADCs in the array will be closely match.

### ACKNOWLEDGMENT

We would like to thank the developers of the Electric VLSI design program at Sun Microsystem, Inc. The Electric VLSI program was used in the design of a test chip.

### REFERENCES

[1]  Dennis Montierth, Using delta-sigma modulation for sensing in CMOS imager, Electrical Engineering Master Thesis, Boise State University, October, 2009.

[2]  R. J. Baker, CMOS: Circuit Design, Layout and Simulation, Revised 2nd ed., Wiley-IEEE, 2007.

[3]  Dennis Montierth, Kuangming Yap, and R. Jacob Baker, "CMOS Image Sensor using delta-sigma modulation," August, 2009.

978-1-4244-6572-9/10 $26.00 © 2010 IEEE

# Poster Presentation Abstracts

## Enhanced Optical Transmission in Hexagonal Plasmonic Crystals

*A. English, L. Lowe, and W. Kuang, Boise State University, Boise, ID, USA*

**Abstract—** The optical properties of periodically modified Ag films are experimentally investigated. These modified Ag films, called plasmonic crystals, are periodically perforated with a two-dimensional hexagonal array of air holes. The transmittance and reflectance of these plasmonic crystals are measured as a function of wavelength and incident angle. It is shown that that surface plasmon resonance in such a hexagonal structures can be described by a Bragg diffraction model. In the model, surface plasmon polaritons are excited by light scattered by one or more reciprocal lattice vectors.

## A Compact Delay-Locked Loop for Multi-Phase Non-Overlapping Clock Generation

*Chris Gagliano and R. Jacob Baker, Boise State University, Boise, ID, USA*

**Abstract —** This paper presents the design, layout, and simulation results for a compact, low-power, low-jitter, delay-locked loop for multi-phase non-overlapping clock generation. A 500 nm CMOS process is used for the design with the total layout size being 810 µm x 95 µm. The operating frequency range is 20 – 100 MHz. The output clock jitter with 400 mV peak-to-peak noise on VDD is 250 ps at 100 MHz. The DLL's outputs consist of eight phases of true and complement, non-overlapping, clock signals (32 total clock signals) buffered to drive standard 8-MOSFET-1-Capacitor switched capacitor circuits. The power dissipation is under 20 mW from a 5 V power supply.

## A 16-bit 500KSps Low Power SAR ADC

*Kun Yang and George S. La Rue, Washington State University, Pullman, WA, USA*

**Abstract -** A 16-bit 500 KSps low power successive approximation register analog-to-digital converter (ADC) in 0.18 µm CMOS process is presented. A self-calibration circuit is implemented to improve the ADC accuracy. An interleaving-by-two architecture is used to obtain a speed of 500 KSps at low power consumption. The power supplies and reference voltages are ±1V and ±0.8V respectively. The ADC power is 4.59 mW including an input buffer and 0.84 mW without. Simulation results show that the signal-to-noise-distortion-ratio, effective number of bits and figure of merit of the ADC are 87.4 dB, 14.22 bits and 44 fJ/conversion step respectively.

## Design of an On-Chip Quasi-Resonant Fixed Frequency Buck DC-DC Power Converter

*Lucas A Wells, University of Idaho, Moscow, ID, USA*

**Abstract—** A quasi-resonant fixed frequency buck DC-DC power converter is designed in American Semiconductor 0.25 µm FLEXFET process. The power converter changes the output voltage using a varactor to modify the resonant response of the buck converter instead of changing or modulating the switching frequency. The effect of on chip passive inductors and capacitors is explored. The test results show that the frequency requirements needed to operate the converter were not achieved. Also the EMI given off by the passive components interferes with the digital control circuitry.

## Method of determination pattern placement errors (PPE) due to scanner lens aberration by using product circuit pattern with double patterning technology

*Maiko Uemura and Masato Shinohara, Micron Japan, Nishiwaki City, Hyogo, Japan*

**Abstract—** We have developed and demonstrated the method to decide the pattern placement errors (PPE) by using double patterning technology for sample making and using inline CDSEM to determine the PPE. In general, contact hole uses conventional illumination condition and line pattern like gate uses annular illumination condition. And there is the delta of overlay result between product circuit pattern and Box-in-Box measurement result due to aberration error which we called PPE. Especially, different type of illumination condition increased PPE. Therefore, the additional offset which we called Non-Zero-Offset (NZO) is required at several process steps. Currently, even if adding NZO, some yield loss is observed due to PPE and slow feedback.

# Poster Presentation Abstracts (Continued)

## Using Gate Voltage Sensitivity to Analyze Bvdss in NAND Periphery RESURF Devices

*Michael A. Smith, Micron Technology Inc, Boise, ID, USA*

**Abstract**—A technique is presented for analyzing the Bvdss path components of NAND high voltage periphery RESURF devices. The Bvdss breakdown path can be determined by the slope of the Bvdss versus gate voltage curve. This methodology allows for optimization of contact spacing rules and provides insight into the effects of process variables on Bvdss.

## Damage Engineering of Boron-Based Low Energy Ion Implantations on USJ Fabrications

*Shu Qin, Y. Jeff Hu, and Allen McTeer, Micron Technology Inc, Boise, ID, USA*

**Abstract**—TEM and SIMS methods and device characteristics are used to successfully study damage engineering of the B-based low energy ion implants. Amorphizing (a-Si) layer (surface damage) and end of range (EOR) damage depths correlate to ion mass (AMU) and implant energy which are scalable to the molecular ion implants. The $^{11}B$ beam-line implant shows thinner both a-Si and EOR depths due to its smaller AMU and lower energy, and the damages are well-annealed under the current anneal conditions. The $BF_2$ beam-line implant shows severe surface lattice damages by over-amorphizing due to its larger AMU and higher energy, and need more thermal budget to anneal (recrystallizing). The $B_2H_6$ PLAD shows moderate a-Si layer and less EOR damages than beam-line ones which attributes to B deposition as a screen so that there are no direct ion bombardments on Si wafer substrate. The $BF_3$ PLAD shows similar damage behavior to the $BF_2$ beam-line implant but with less lattice damages. The device performance evaluation confirms that the device processed by $B_2H_6$ PLAD shows better $I_{ON}$ versus $I_{OFF}$ characteristics than those processed by $BF_2$ and $C_2B_{10}H_{12}$ molecular beam-line implants due to less damages and well-annealing.

# IEEE WMED 2011
# Advance Call for Papers

Ninth Regional Meeting of the Workshop on Microelectronics and Electron Devices, Spring 2011

CALL FOR PAPERS

The ninth annual IEEE Workshop on Microelectronics and Electron Devices (WMED-2011) will provide a forum for reviewing and discussing all aspects of micro- and nano-electronics including processing, electrical characterization, design, and new device technologies. This symposium will consist of invited and contributed talks, papers, and a poster session throughout the day. Faculty, students, and researchers in industry are encouraged to contribute presentations on either completed research or works in progress. Topics in the following areas will form the contributing sessions and poster session in the workshop:

MICROELECTRONIC DEVICE PROCESSING AND PROCESS INTEGRATION

Trends in submicron technologies; product development (DRAM, SRAM, Flash, CMOS Imagers); new device technologies (Phase Change Memory, Resistive Memory, Ferroelectric Memory, Nano-electronics), novel transistors.

MICROELECTRONIC DEVICE, CIRCUIT DESIGN, AND RELIABILITY TESTING

Dielectric reliability; device reliability; phase change memory reliability; novel memory technology testing schemes; electrical properties of novel devices. New product design, design techniques and memory sensing schemes.

SEMICONDUCTOR PACKAGING AND RELIABILITY

Semiconductor package reliability, Design for Manufacturability, and stacked die packaging, and novel assembly processes. Novel packaging structures, processes, research, development, and performance.

MEMS AND NANOELECTRONIC DEVICES

Novel processes and materials, MEMS research, development and performance; nanotubes, nanowires, quantum dots, molecular devices, device characterization for nanoelectronic devices.

An IEEE Publication of the accepted papers and a CD-ROM including the papers and presentations is planned and will be available at the start of the workshop. Submitted manuscripts must follow the IEEE publication format guidelines.

See the WMED website or additional details:
http://www.ewh.ieee.org/r6/boise/eds/WMED.html

This workshop is receiving technical co-sponsorship support from the IEEE Electron Devices Society.

978-1-4244-6572-9/10 $26.00 © 2010 IEEE

9781424465729